CONSUMER HANDBOOK OF SOLAR ENERGY

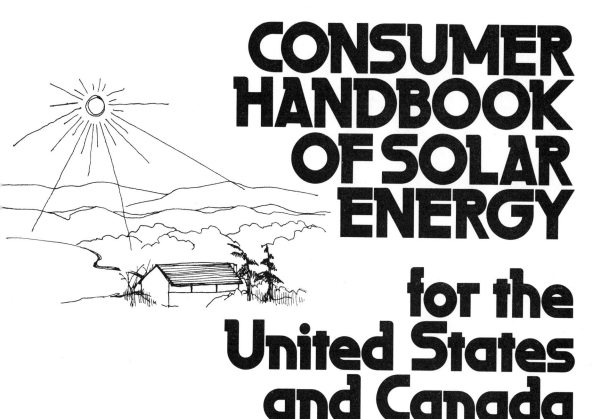

CONSUMER HANDBOOK OF SOLAR ENERGY

for the United States and Canada

JOHN H. KEYES

MORGAN & MORGAN
Dobbs Ferry, New York

Earth Books
a division of
Morgan & Morgan, Inc.,
145 Palisade Street
Dobbs Ferry, New York 10522

International Standard Book Number   Paper 0-87100-149-7
                                    Cloth 0-87100-156-X

Library of Congress Catalog Card Number 78-65084

Printed in the United States of America

Book design and typesetting by Eagle Spring Press
Illustrated by Linda Schifano

DEDICATION
This book is gratefully dedicated to wife Erika, who
has the incredible ability to live through the mess
and moods without murmur; to son John-Michael and
daughter Heidi-Rika, who should complain, but don't;
to Betty, whose loyal certainty and prayers have been
appreciated, and to the special friends who have believed
since the beginning.

FOREWORD

I am very excited about this Handbook. Never before has an innovative technology required so much of the buyer. Never before has so much mythology accompanied the introduction of a new product into a marketplace.

Buying a pig in a poke has been the subject of untold numbers of jokes since the Middle Ages, yet today the manufacturers of much of the solar equipment on the market are requiring their customers to do just that.

Rather than telling their customers the performance characteristics of their equipment, they publish irrelevant "collector efficiency graphs" it takes an engineer to read and, even if one reads the graphs, they don't tell how the collector works under ordinary, average conditions in your home town. More important, and this is the scandal of the infant solar energy industry, most of them will not tell you what the heat output from their equipment will be *inside your home*, nor do they provide any output information for your community. That is important because the very same solar furnace does not have the same Btu output (British thermal units, a way of measuring heat) in different towns. Obvious, if you think about it. The amount of cloudiness in your home

town is different from that of a town 50 miles away. The winter conditions are different, the temperatures are different. Those all make a difference to the performance of a solar furnace.

Even some large companies which are now getting into the solar-energy field expect you to buy their equipment sight unseen, at least from an output standpoint. For example, one of these has a statement in the sales brochures: "It takes a collector from 15% to 30% of the square footage of your home to heat it." That brochure goes to homeowners in Arizona and in Caribou, Maine. Obviously the furnace sizes required for the very same house would be quite different, and not by just 15%! Even in the same town, two houses with the same square footage, sitting side by side but having differing amounts of insulation, would require different sizes of solar furnaces.

Surely the engineers at that large company know better, but you shouldn't have to be an engineer to buy solar equipment.

Because of the widespread consumer abuse in sales of solar units, this Handbook can be very useful to you. I am sure it will save you hundreds of times the cost of the book when it comes time for you to purchase equipment.

For your convenience, a partial list of deceptive practices I have observed in the industry is included in Appendix G, and the rules to follow while shopping for solar equipment are summarized in a convenient checklist in Appendix J. These lists appear without accompanying explanations. To understand the reasons for inclusion of various items, read Sections One, Two and Three.

This book reveals, for the first time anywhere, the methods for comparing various solar furnaces on the same basis. Even if the manufacturer is not shooting square with you, you will know what the approximate heat delivery from that system will be, in your home town, inside your home.

Solar energy is not a panacea. You will be able to determine whether or not it will pay for you to put a solar collector on your home. And in many cases, you will find that it simply is not economical for you to do so.

Because I will be hammering at some of the practices in this new field, you need to know my orientation, what axes I am grinding and what my hidden interests are.

A number of manufacturers—*in competition with each other*—produce solar equipment under patents which are licensed by a small, privately funded research corporation owned by my wife and myself. I have deliberately avoided mentioning the names of those manufacturers or their trade names. They have brought the price of the equipment down by nearly 50% in the last five years, and, in most cases, improved the design that I patented. They pay nominal royalties to our corporation. As of this writing, they account for nearly 90% of all the solar-heated buildings in the United States and Canada. Some of them are guilty of marketing practices that I will be hollering about in this Handbook. I hope that when you have finished reading you won't know which units are of my design. I have sincerely tried to avoid pounding any one drum.

Because, despite all the rascals in the new solar energy business and all the rapacious practices, they are far outnumbered by the good, idealistic folks who are really trying to bring you good products at reasonable prices. There are lots of good solar furnaces and lots of good solar-powered hot-water air heaters to be had.

I must apologize to my Canadian friends. The Degree Days listed are Fahrenheit Degree Days, and the heating outputs are given in Btu's instead of in calories. (See Chapter Ten for the definition of a Degree Day.) It was just too confusing to mix Celsius and Fahrenheit and Btu's and calories in the same tables and charts. Conversion charts are provided for your use in Appendix F.

The astute reader may notice some differences in the solar-furnace Btu outputs listed for Sault Ste. Marie, Michigan, and Sault Ste. Marie, Ontario; for Port Huron, Michigan, and Sarnia, Ontario; etc. These differences can be attributed to the difference in location of the measuring stations and to slightly different measuring techniques used by the United States and the Canadian weather bureaus.

All the 70,000 or so calculations needed to prepare this Handbook were done by me; so any errors can be laid at my doorstep. Obviously, I have done everything possible to eliminate all errors, but many are sure to survive simply by virtue of the quantity of material. It is just for that reason—the number of calculations needed—that most of the material presented in this Handbook appears for the first time anywhere.

John H. Keyes
January, 1979
Nederland, Colorado

CONTENTS

NOMOGRAMS

FIGURES

TABLES

INTRODUCTION

HOW TO USE THIS HANDBOOK

This Handbook has been organized into eight separate parts for your convenience. Depending upon what your present interests are, you can refer directly to the section that is of most interest to you.

I hope you will have the time and interest to go straight through the Handbook. If, however, you are not planning to buy solar equipment immediately, Section Two will be of most interest to you because it will show you how to convert your home into a kind of energy-conserving passive solar heating system. Even if you are poor, or on a fixed income, it is possible to cut your heating bill in half with as little an investment as $25 to $50. You will even find instructions there for making your own insulation so inexpensively that you can insulate a small house for less than $25!

Section One consists of an introduction to solar equipment and a comparison of the advantages—and disadvantages—of various kinds of solar furnaces. Of particular interest in Section One are the guidelines for purchasing solar heating and hot-water equipment. You will be able to spot the cheats a mile away. With your own good, common sense you can determine whether it is better to purchase a water collector, an air collector or a new high-technology unit. Simply following these rules (reviewed in Appendix L) can save you as much as $15,000 on the price of a solar furnace for your home.

Section Three provides chapters on selecting the best dealer to install your solar equipment and tips on purchasing solar energy stocks. It also provides, for the first time anywhere, simplified multipliers for determining the Degree-Day size of your house —whether it is of wood-frame, brick, concrete, cinder-block or log construction—using just a pencil and a tape measure and about half an hour of your time. Simplified worksheets have been provided for your convenience, as well as full instructions and sample calculations.

Section Four contains special nomograms that will tell you how large a solar collector your house, in your home town, needs. These are called Solar-Collector-Grams. All you need is a ruler or a straightedge and a piece of paper to find the number of square feet a solar collector needs to have to heat your house, and the size needed to heat the water your family uses. These Solar-Collector-Grams represent a real advance in simplifying what previously has been a day-long job for an engineer. Now for the first time you the consumer can tell at a glance just how much equipment is needed to supply from 40% to 90% of your heating requirements, so that you can make a pocket-book decision.

Section Six contains other nomograms, ones called Solar-Pay-Back-Grams. All you need is last month's utility bill and the cost per square foot of solar collector to find out —using just a ruler or straightedge—just how long it will take you to pay for your solar equipment in fuel savings, whether you are using natural gas, propane, fuel oil or electricity to heat with now.

Section Five is designed to allow you to outfox the manufacturer or salesperson who won't tell you critical information about the equipment that he or she is selling. Using special tables and charts, with worksheets provided for your convenience, you can calculate the Btu output *inside your home* of any flat-plate or vertical-vane solar collector sold on the North American continent.

Section Seven contains the ranking of some 342 U.S. and Canadian cities by how much energy the very same solar furnace would provide to a consumer in each of those cities. You will find a few surprises there. For example, a solar furnace in San Francisco can be smaller than one on the same house would be in Dallas, Texas, or Albuquerque, New Mexico. Washington, D.C., ranks higher than Denver, Colorado! This ranking should help to destroy the myth that solar equipment is only practical in some "sun belt".

Finally, there are a number of Appendices for the use of those who wish to delve more deeply.

Every effort has been made to make this Handbook clear, concise and understandable to the layman. Many myths are destroyed. The explanations refer, wherever possible, to things you are already familiar with, so that you can evaluate them by your own experience and common sense.

Unfortunately, as for all emerging technologies, great amounts of mystery and misinformation surround the use of solar energy.

For the technical person desiring to understand how certain tables and charts were derived, notes are provided at the end of those materials. In addition, some of the tables used to derive Btu outputs are included in the Appendices for your information.

SECTION ONE

An Introduction to Solar Energy Equipment

1 Flat-Plate Solar Collectors

HOW THEY WORK

For a number of reasons, which will become evident later, the solar collector you use on your house will most likely be a *flat-plate collector*. Unless you are filthy rich and plain foolish. Or unless you are getting a government grant.

Why? Because for home solar heating applications the flat-plate collector is much less expensive than the other kinds of solar collectors.

Therefore, it is to your advantage to understand the design and operation of the flat-plate collector fairly well. Fortunately that isn't difficult, since the flat-plate collector is simply an insulated box with a clear cover designed to trap sunlight.

It really is very simple. Even though there are literally hundreds of variations on the flat-plate collector, the principles of operation are the same for all of them and, surprisingly, if they are constructed with the same amount of insulation they all work with about the same efficiency, despite their manufacturers' claims to the contrary!

CONSTRUCTION OF THE FLAT-PLATE COLLECTOR

The flat-plate collector consists of three basic parts: (1) an insulated housing, (2) a transparent cover or covers made of glass, glass fiber or plastic and (3) an absorber plate made of aluminum, cloth, copper, glass fiber, iron, plastic or screening. The absorber plate is typically coated or painted with black or a dark color.

THE VERTICAL-VANE COLLECTOR

Some flat-plate collectors have a variation called the "vertical vane." This is only a variation and does not change the category of a flat-plate collector. The vertical vane is simply a device used to increase the area of the absorber plate inside the collector.

This can be done by a number of methods, any of which is approximately as useful as any other. Most common are: the "beer-can collector," which uses a series of shallow aluminum cups, giving the collector a honeycombed appearance; vertical or horizontal fins resembling a venetian blind in general

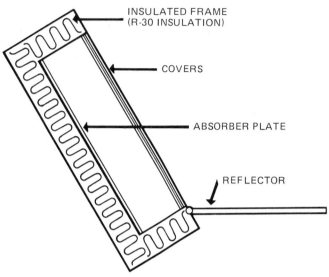

FIGURE 1: Construction of the typical flat-plate collector.

appearance; stampings, which dimple the surface or raise small fins from the absorber plate; etc.

There are very good reasons for increasing the area of the absorber plate. One is to improve the rate of heat transfer from the collector to the storage and another is to combat the mortal enemy of solar collectors: dust.

But don't be deceived. That increased surface area inside the collector does not allow a smaller collector to *receive* more energy than a larger one without the vertical vanes. Only so much energy from the sun falls on a square foot of a solar collector's cover. If two collectors have the same number, kind and square footage of cover, and are sitting side by side, the same amount of energy will get inside the collectors regardless of the configuration of the absorber plates. However, the vertical-vane collector will *lose* slightly less energy back out of the collector than a flat absorber will. The energy is bounced around more *inside* the collector, therefore it will collect more energy than a flat absorber. Not because of the increased area. Rather, because of the shape of the absorber, the amount of energy lost from the collector by *reflection* is reduced.

REFLECTION ENERGY LOSS

A good flat black paint will absorb about 95% of the incoming light if the light is perpendicular to the painted surface. That means that about 5% is reflected away. If the incoming light is at an angle to the surface, even more is reflected. At extreme angles (like sunlight in the early morning and late afternoon), as much as a third of that light is reflected away. To see how that works, look straight down at the black square below. It will look very black. But if you hold it at an angle in front of a window, it becomes a fairly good reflector and will appear to the eye almost white.

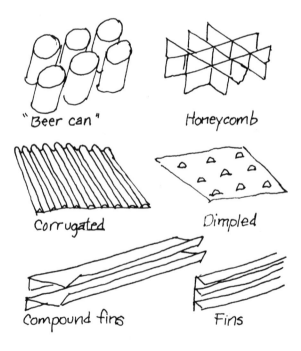

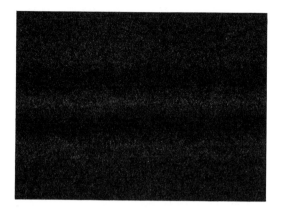

FIGURE 2: From directly above, this black square looks very black, but at oblique angles (in front of a good light source) it looks white.

Thus, when the sun is at extremely oblique angles to the flat-plate collector, that 33% reflectivity becomes important because nearly that amount of energy is bounced right back out of the collector without being absorbed, converted to heat and transferred to storage.

To see how a vertical-vane-type collector changes this, look at Figure 3, below. As you can see from the diagram, the incoming energy is bounced back and forth inside the collector *before* it gets the chance to be reflected back through the covers.

Let's assume that 33% of the incoming light is being reflected. On the first reflection, 33% is not absorbed by the black surface but continues on to the second surface. On that reflection, 33% of that 33%, or 11%, is not absorbed, but reflected onwards. On the third bounce, again only 33% is reflected. But that is 33% of 11%, or about 3½%. So, even with extreme reflection, less than 3½% is lost back out of the collector through the covers.

In actual practice, the vertical-vane collector's loss due to reflection is less than 1%

over the course of a day. The flat-plate absorber will lose about 5% because of reflection over the same period of time, when the collector is new.

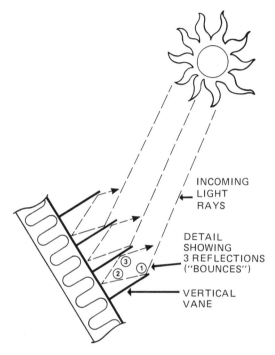

INCOMING
LIGHT
RAYS

DETAIL
SHOWING
3 REFLECTIONS
("BOUNCES")

VERTICAL
VANE

FIGURE 3: Diagram of the vertical-vane absorber plate, showing the reflection of energy inside the collector.

It won't hurt to repeat the fact that no more energy *gets into* a vertical-vane collector. It simply *stays* there.

Now, however, let's get to the prime rule concerning flat-plate and vertical-vane flat-plate collectors.

RULE NUMBER ONE

If insulated alike, all flat-plate collectors with the same number and kind of covers work the same when new. Vertical vanes improve performance about 4% when new.

Salespeople who tell you otherwise are lying to you. If one tells you that in some mysterious way his or her collector is superior to other flat-plate collectors, you should start muttering about walking in barnyard carpeting and start for the door. It is always risky to do business with liars.

COLLECTOR COVERS

When dealing with collectors, you may encounter some fast talk about "high transmissivity" covers. Be careful. Two ordinary tempered glass panes (like your sliding glass patio doors) will only permit about 77% of the incoming sunlight to get inside the collector. So it would seem that a tremendous improvement could be made simply by improving the clearness of the glass, allowing more energy to get inside the collector to be collected.

It ain't quite that easy. To understand why, you must know where the energy goes that doesn't get inside the solar collector. Where the surface of the glass meets air there is something called an "air/glass interface". One pane of glass has two air/glass interfaces, two panes have four. Each air/glass interface reflects away 4% of the incoming solar energy. So that, if you have two panes of glass, four times 4%, or a total of 16%, of the incoming light will be reflected away and never get inside the collector. (Actually, only 15% is reflected away, because 96% of the light passes the first air/glass interface, 96% of 96% passes the second, and so on.)

Now, if you improve the clarity of the glass, that is, buy a "low-iron" or "water-white" glass, you are not going to help the *reflectivity* problem. Two panes of clearer glass will still have four air/glass interfaces to reflect the energy away. You are still going to lose about 15%.

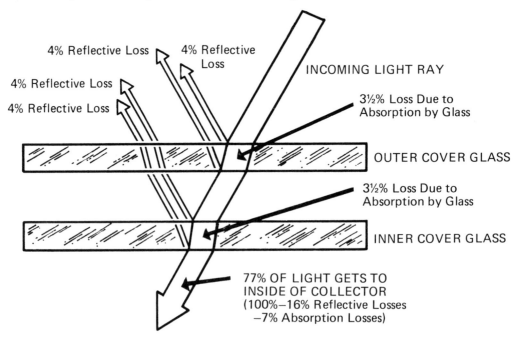

FIGURE 4: Four per cent of incoming energy is reflected back by each air/glass interface.

The cheaper high-iron-content glass absorbs about 4% of the incoming energy for each pane. With two panes of glass, energy absorbed by the outer glass cover is lost, but energy absorbed by the inner cover contributes to collector efficiency. So the best improvement possible, utilizing "low-iron" or "water-white" glass, is only 3%.

And "low-iron" or "water-white" glass is definitely going to cost you more than ordinary glass. I believe that there is some question as to whether or not the additional cost is justified. It can be argued both ways. However, it most certainly will take a number of years to pay off the additional cost with the slightly improved performance. From an over-all net energy perspective: It costs more energy to purify the clearer glass, and that means that the collector must work longer to pay back the energy it took to make it. If you are willing to pay the additional price for the slight improvement in performance, that's your business.

I wouldn't turn cartwheels in the solar-furnace showroom, however, if the salesperson solemnly tells you that the higher priced "low-iron" or "water-white" glass is being used in the collector. Regardless of the type of glass being used, be very sure that it is *tempered* glass.

ANTIREFLECTIVE COATINGS

There are special antireflective coatings that can be applied to collector covers to make them less reflective, to attack directly the problem of high losses. Unfortunately, at this time they are still very expensive and they do not have a good life expectancy. If a dependable, long-lived, inexpensive antireflective coating is developed, it will undoubtedly be adopted by all manufacturers of solar-power equipment, since it will improve collection very significantly.

THE GREENHOUSE-EFFECT MYTH

There is one thing you should understand about the solar collector before we go any further. Old-time hobbyists and early investigators of solar-energy equipment (myself included) thought that the glass covers played an important part in something called "the greenhouse effect."

Here's how the myth went: Solar energy in the 0.25 micron to 2.5 micron wavelengths (called *shortwave* solar energy) passed through the covers and was absorbed by the plate. This energy was "converted" into energy in the 3.0 micron to 30 micron wavelength (called *longwave* radiation). Because the glass covers were transparent (relatively) to shortwave radiation, but nearly opaque to longwave radiation, the energy was "trapped" inside the collector. This was called "the greenhouse effect."

As we all learned more about this process, we found out that it was really a lot simpler than that. If the heat is being removed from the solar collector efficiently with air or water, and if the system is a properly engineered low-temperature system, there really isn't much "longwave radiation" taking place. (In general, it takes a pretty big temperature difference to get significant amounts of radiation.) The main function of the covers turns out to be the prevention of convective (air movement) and conductive (contact transfer) heat losses from the collector during the time the heat is being moved by fan or pump to storage.

So, if your salesperson talks about the "greenhouse effect," or, worse yet, if the manufacturer's literature does, you can be sure that you are dealing with neophytes using outdated information.

COLORS OF COLLECTOR SURFACES

As we saw earlier, black absorbs more light than other colors, and so most absorber plates are painted black. But in the real world, dark red, green, brown or blue will absorb nearly as much light as black. Therefore, if you see a collector painted one of those colors don't worry about it, particularly if it is of the vertical-vane type. The different-colored collector will work as well, for all practical purposes, as the black one. The only time you should get worried is if the salesperson tells you it works *better!*

SELECTIVE COATINGS

Probably one of the funniest bits of pseudo-scientific high-technology mumbo jumbo to emerge from the infant solar industry is something called the "thermally selective coating." This "development" grew out of the myth of the "greenhouse effect." Some very smart Israeli scientists came up with a coating for absorbers that would absorb shortwave radiation but would not re-radiate longwave radiation. I am not demeaning the importance of this discovery in high-temperature applications where radiative heat loss can be a real problem, such as in space. But here on Earth, in a properly designed low-temperature solar collector for residential heating, radiative heat loss is not a big problem.

And those manufacturers who use selective coatings in their collectors overlooked a second problem: All collectors get dusty. Unless they are perfectly sealed. And I've not ever had the good fortune to see a flat-plate collector that was perfectly sealed. We'll discuss why not in the section on sealants and glazing.

This dust presents a big problem to the selective coating, even if it is inside a high-temperature collector where it might do some good. If the selective coating does not radiate energy in the 3.0 micron to 30 micron wavelengths, neither does it *absorb* energy in that spectrum. When a layer of dust

accumulates, that layer of dust becomes an absorber plate. The *dust* absorbs some of the shortwave energy coming through the covers and, in a high-temperature application, re-radiates that energy in the longwave spectrum. If the absorber plate has a selective coating, the system has been cunningly designed so that it *cannot* absorb that re-radiated energy!

So, out here in the real world, away from the neat and tidy white lab of modern science, collectors get dusty, and over a year or two the ones with selective coatings will degrade in performance much more than a simple flat-black-painted collector will. Therefore, if some enthusiastic salesperson starts waxing poetic about the advantages of selective coatings in home solar heating systems, don't be taken in. If it is a properly designed low-temperature system, the selective coating only adds to the cost of the system, doesn't perform a bit better and, in fact, will perform worse over a period of time. (Remember Rule One: All flat-plate collectors perform the same, if insulated the same, when new.)

THE INSULATION OF COLLECTORS

The next rule is the quickest single way for you to spot incompetent design in a solar collector.

RULE NUMBER 2

A decently designed collector should have at least as much insulation on its sides and back as you have in the attic of your house.

Why is this rule true? Simple. *The collector never turns on unless it is warmer than you keep your house.* Use your common sense. If the collector only works at higher temperatures than you keep your house, it only stands to reason that it needs at least as

much insulation as you have in your house.

I recently had a discussion with an engineer new to solar-energy applications, who maintained that because solar energy is free one should not insulate collectors at all. That is, of course, downright stupid. The poorer the insulation on the collector, the bigger it has to be to do the same job.

Before we go any further, you should understand some jargon. When talking about insulation, "R-Values" are used. The "R" stands for "Resistance to heat flow." The bigger the R-Value, the better the insulation.

Let's take an example to see just how big a difference insulation can make. A solar heating system with two inches of glass-fiber insulation (R-6.67) on collector and storage would have to have 690 square feet of collector to equal the performance of the same system with ten inches of glass-fiber insulation (R-30) on collector and storage, but with only 150 square feet of collector. Why is that important? At today's prices, you can expect to pay between $20 and $60 per square foot of collector for the entire system, installed. Even at $20 a square foot, that is a *difference* of $10,800. Insulation is very important to your hip pocket when it comes to solar equipment.

On the other hand, from the standpoint of some manufacturers, if the system is poorly insulated the sale to you is much larger, the factories stay busier, more jobs are provided and, incidentally, manufacturing profits are larger. My opinion is that such manufacturers are stealing from you as surely as any mugger does.

In Chapter Thirteen, on pages 131 through 139, appear the tables for percentage efficiencies of collectors and storage units with various amounts of insulation. A survey of the market, however, shows that you will pay just as much for rotten collectors with totally inadequate amounts of insulation as you will for ones with appropriate design and adequate amounts of insulation. One with R-3.33 or R-6.67 insulation (one or two inches of glass fiber) retails for as much per square foot as one with an

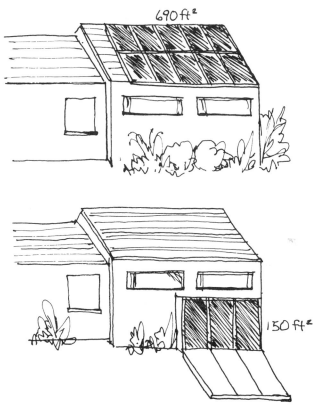

690 ft²

150 ft²

R-15 or R-30 (two or four inches of foam). Therefore, seek out the very best-insulated collector for your home when you start shopping for solar equipment. The extra insulation isn't going to cost much, if any, more.

COLLECTOR COVERS

One of the biggest potential problem areas on your solar collector is the cover or covers.

RULE NUMBER 3

Unless you live at a latitude of less than 30 degrees North, don't buy a flat-plate collector with fewer than two covers. The covers should be of tempered glass. If glass fiber or plastic is used for a cover, it should be the outer cover only, with the under cover of tempered glass.

This rule may change slightly in the future, if some new developments provide fibrous glass or plastics that have higher temperature capabilities.

The temperature in a properly insulated collector, even if provided with automatic, self-actuating venting devices, can climb to 375° Fahrenheit in the event of a power failure on a warm sunny day or in the event of any mechanical problem that stops the flow of air or water through the collector. Yet none of the materials being sold for collector covers, except glass, can withstand that temperature without degrading. At least, all the glass-fiber and plastic materials that I have tested have failed to withstand 375°F for two hours. Some turn bright yellow or orange. Some crack, some get brittle. Some melt.

Since it is fairly certain that a power outage will occur sometime during the life of your solar equipment, it is stupid to gamble that you won't get stuck with the replacement problem. *Particularly since no manufacturers, to my knowledge, provide warranties on the covers.* In fact, most of the warranties I have read specifically state that the covers are excluded. Even if the damage or breakage is due to the manufacturer's rotten design!

TEMPERED GLASS VERSUS PLASTIC

If the glass covers are not tempered, you can be fairly sure that you will be replacing them some day. One of the reasons that a lot of people like glass-fiber or plastic covers on their collector is that they are worried about breakage by vandals, or by hail or wind.

It is a fact, however, that a plastic cover is not a bit more resistant to a determined attack by a vandal than is tempered glass. If there is a vandal who wants to destroy some of your property, it will happen. The collector cover is certainly no more susceptible to damage than the windows in your house. And it takes more of a hailstorm or windstorm than you are likely ever to experience to break tempered glass. I have seen tempered-glass covers survive winds in excess of 75 miles per hour and hailstones that averaged about 3/4 inch in diameter.

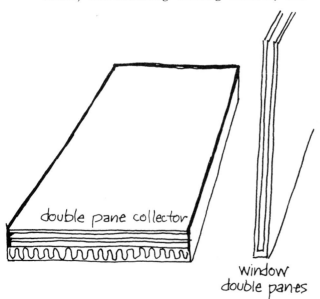

double pane collector

window
double panes

I think that all the above facts are known by most manufacturers. Probably the key reason for using glass fiber or plastic is cost. Tempered glass costs a lot more. Again, however, the more shoddily constructed collectors *don't seem to sell for less* than the ones that have been built with you in mind. Whatever savings are realized just end up in the manufacturer's pocket.

RIGIDLY BONDED GLASS

It is quite possible that you have thermal-pane windows in your home. These may very well consist of two panes of glass actually bonded to one another rigidly with an encapsulated air gap. For windows, this type of construction will work well.

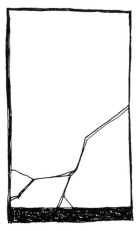

Unfortunately, such rigidly bonded panes will *not* work on a collector. In a collector, the difference in temperature between the inner pane (which is at the temperature of the absorber plate) and the outer pane (which is near the outside temperature) can be quite large. The result, if one pane is expanding while the other is contracting, is quite obvious: One or both will break. The proper design of collector covers must provide for independent expansion and contraction of the two glass covers.

RULE NUMBER 4

Don't buy a collector that has the two glass covers bonded rigidly to each other.

CONDENSATION BETWEEN COVERS

Unfortunately, providing for independent expansion and contraction of the two covers can cause other problems. If the two panes are not sealed perfectly, condensation can occur between them, reducing the effectiveness of collection.

One possible answer is to put some kind of dessicant (a moisture-absorbing chemical) between the panes. Unfortunately, in practice this solution doesn't always work very well.

If one has an air/solar system, the usual solution is to leave the inner pane unglazed or to provide small holes, which allow a limited amount of flow of air between the panes. This small flow tends to keep the condensation from becoming a problem. If the flow between the panes is too great, however, the benefits of having two covers are lost.

In water/solar systems, some other method must be used.

RULE NUMBER 5

Ask about the method used by the manufacturer to prevent moisture build-up between the collector covers. Then evaluate it with your own common sense.

THICKNESS OF GLASS ON COLLECTORS

Take a look at the glass covers on the collector you are planning to buy. Although the panes may be as large as 4 feet by 8 feet in size, you will find glass supports inside the collector. If the glass spans a distance of 2 feet without support, it should be at least

3/16 inch thick. If it doesn't span more than 18 inches, 1/8 inch-thick glass will probably suffice.

You may be thinking by now that all this talk about covers is unnecessarily long and boring, *but the covers can be and probably will be the single most troublesome part of your solar furnace.* That is *exactly* why the manufacturers exclude the covers from their warranty coverage!

PAINT

The next thing to worry about is paint. This one step can save you a ton of headaches. Why? Because, in the rush to market, many of the manufacturers haven't done their homework as well as they should have.

RULE NUMBER 6

Insist upon seeing the results of a seven-day stagnation test performed by an independent testing laboratory on the collector you plan to buy. (Sometimes this is called a "stasis test.")

If the temperature in the collector gets up into the higher ranges, something called "outgassing" occurs, usually from the paint used on the absorber and generally because a specially formulated high-temperature paint wasn't used by the manufacturer. The paint will outgas small particles onto the under cover, giving it a fogged appearance. Naturally, this fog impairs the performance of the system because it stops some of the sunlight from entering the collector. (Other materials, such as sealants or plywood, can also outgas.)

The stagnation, or stasis, test is extremely simple. The collector is put in the sunshine (usually in Arizona, Southern California or Florida) with no heat removal taking place. After at least seven days (some

labs do a 30-day test), if the outgassing is going to occur, there will be evidence of it.

The reason this is so important may not be apparent to you. After all, all you would have to do is open up the collector and wash the cover, right? Nope. It is extremely difficult to deglaze a collector cover. If you don't do it yourself, it will cost a real bundle of money. Yet if the manufacturers had done their job correctly, there never would have been the need. So, if a manufacturer isn't willing to provide you with the results of independent testing to prove that the collector won't outgas, go somewhere else. To a reputable manufacturer. Believe me, you don't want the grief of cleaning a collector!

SEALANTS AND GLAZING

It is nearly impossible to seal a collector perfectly. There is a relatively great amount of expansion and contraction taking place every day in a solar collector. This expansion and contraction occur at different rates in different materials. That is, glass doesn't expand and contract at the same rate as aluminum, wood or steel. The collector gets hot in the daytime, but it gets just as cold as anything else out of doors during the nighttime.

Unfortunately, no sealant has the "memory" to "remember" such changes and to handle such movement on a continuing basis over the lifetime of the collector.

At some point, whether one year or five years from now, it will be necessary for you to reseal the collector covers. Sometimes this procedure is only moderately difficult, sometimes it is very difficult. It is always a pain in the lower back.

RULE NUMBER 7

Be sure to find out what kind of sealants have been used on the collector. Then look

at the collector yourself to see just how difficult it will be to recaulk the covers after a year or two.

Sealant questions are fun, since they really give you a chance to size up the salespersons you are dealing with. If they are evasive, look out. Have them show you what is involved in recaulking the collector covers. If it sounds easy, something is wrong, because it *never* is.

DUST
Some of the major horror stories in the infant solar-energy industry relate to dust. There were projects where the huge rooftop arrays put in place with governmental funding had enough dust build-up to require costly cleaning of the absorber plates. This negative happening is hushed up, so that it is a bit difficult to find out just how much the cleaning jobs cost; but rumors abound of universities spending as much as $10,000 just to clean up their solar projects.

RULE NUMBER 8
Make sure that the manufacturer has addressed the dust problem. Be very sure that the buck hasn't been passed to you.

Dust is a problem hardly anyone in the solar industry wants to talk about. As we have seen, it is nigh unto impossible to seal a collector perfectly. In operation, whether it is an air-type or a water-type collector, it operates somewhat like a giant vacuum cleaner. That is, as the absorber heats up, the air becomes less dense and air is expanded out of the collector. But at night the reverse occurs as the collector cools to the same temperature as the air outside, and dust may be drawn in with the contracting air. Over a period of time, a fine, almost invisible layer of dust forms on the absorber plate.

After just a year, a significant drop in collection efficiency can be observed. Why does this happen? The dust increases the reflectivity of the absorber plate, which means that more energy is reflected back out of the collector instead of being absorbed and transferred to storage.

HINGED COVERS
What is the solution to the dust problem? A couple of methods are practical.

One is to provide special hinged collector covers with pressure-type seals. This solution faces up to the problem squarely. It assumes that you are going to have to clean the collectors on a regular basis and provides a reasonable method for making that job somewhat easier.

Obviously, if you are purchasing a collector with hinged covers and pressure seals you need to examine the seal material carefully. See if you can examine a similar collector that has been in operation for at least one year. Open up a panel and check *those* seals. This simple procedure may save you a lot of grief later.

VERTICAL VANES
A second solution to the dust problem is the vertical-vane flat-plate collector, which con-

tinues to perform extremely efficiently even when it gets very dusty. Mother Nature is made use of to "solve" the problem—by accommodating to it.

Here's how it works: As we saw earlier, the vertical-vane configuration bounces the incoming light around inside the collector. The vertical vanes get dusty, just the way the flat-plate absorber does. But as the reflectivity increases from the original 5% to 15% or 20%, the performance of the vertical-vane collector doesn't decline by the same amount as the flat absorber's does. Instead of declining by 15% to 20%, it only declines by about 1%.

In other words, the vertical vanes permit the collector to live with, to accommodate to, the dust without a significant decline in performance. After five years, however, the dust will be visible when you look into the collector. This may bother you for aesthetic reasons, even though it is still working nearly as well.

CLEANING COLLECTORS
The reason that cleaning the collector is so to be feared, is simple. The sealants or glazing materials used to seal the collector have

had a long time to "cook" and to glue the collector together. When you clean the collector, you have to break those seals without breaking the glass, which is a heck of a lot harder to do than you might think.

Then you have to lift that glass out of the collector. A 4-foot-by-8-foot pane of tempered glass may well weigh more than 150 pounds. Begin to see the problem?

Then, just for an exercise in foolishness, imagine yourself wrestling that huge pane of glass around on a 60°-angled roof! Only a college professor working in an ivory tower far removed from the real world, or a manufacturer who doesn't care, would dream of putting you in that position.

Nearly 90% of all the solar-heated homes in the United States today have their collectors mounted at ground level, instead of on rooftops. This begins to make a lot of sense, doesn't it?

COLLECTORS ON ROOFTOPS
That brings us to looking at the location of your solar collector. Where should it go? I know, I know. There's been so much publicity showing an entire rooftop filled with collectors. Listen. Those were *government* installations. That's right. The same people who run the social security program so efficiently, and balance our national budget every year, also get involved in solar-power research. THE ROOF IS A GOOFY PLACE TO PUT THE COLLECTOR. Only someone who has absolutely no other place available to put the collector, or a nut, would put a solar collector on the rooftop. Oh, I'll freely admit that it *is* 10 feet to 20 feet closer to the sun up there. I suspect that that is exactly the reason why the first solar collectors were put on the roof. But in 93,000,000 miles (the distance to the sun), 10 feet only gives you a 0.000000002% advantage!

In actuality, the collector doesn't work better on the roof. *It works worse.* For one thing, it is slightly colder up there and more exposed to the wind; so the heat loss from the collector is a bit higher, meaning that the efficiency is lower.

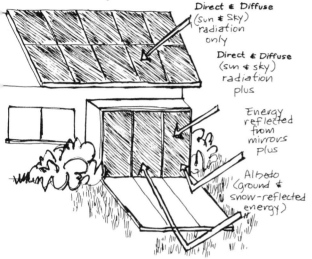

Direct & Diffuse (Sun & Sky) radiation only

Direct & Diffuse (sun & sky) radiation plus

Energy reflected from mirrors plus

Albedo (ground & snow-reflected energy)

A second reason it is wacky to put the collector on the roof is this: It will cost you anywhere from $2,000 to $6,000 extra to frame up the roof at a 60° angle to accommodate the thing there and to handle the weight and the wind loads safely.

Finally, the collector mounted on a rooftop doesn't get *as much* solar energy as one mounted at ground level. That's right! *You will get from 5% to 30% additional solar energy in a ground-mounted collector.*

The thing to remember is this: Your collector is a *light-trap*, collecting light from the sun. On the roof it will receive direct radiation and diffuse sky radiation. But it will be missing an important source of light. A collector mounted at ground level will get both the direct and the diffuse inputs that a rooftop unit will, and in addition it will receive something called "albedo inputs." "Albedo" is a fancy word for ground-reflected energy! Solar energy that is bounced into the collector off the grass, the snow, neighboring buildings, etc.

Measurements have shown that plain, ordinary, clean snow will reflect as much as 87% of the energy striking it. Grass reflects about 25%, concrete reflects about 30% and, surprisingly, even a blacktop parking lot will reflect 10%.

It is important to notice that snow reflects 87% of the energy striking it. This reflection is not the neat, orderly kind you get from a mirror. If you look down at snow, you can't see your reflection. The light is not being reflected in parallel rows from the snow. Rather, it is scattered, or *diffused. A sunlight collector doesn't care.* The absorber will absorb this diffused light just as handily as it will any other light getting inside the collector. As we will see later, in the section on mirrors and reflectors,

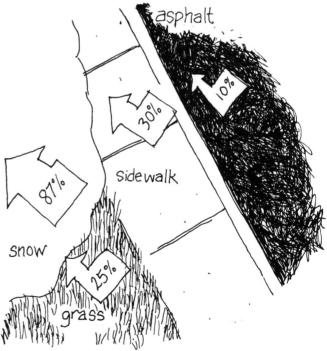

asphalt

30%

10%

sidewalk

87%

snow

25%

grass

many early scientists made the mistake of thinking that only the neat, orderly reflection was important. It turns out, for example, that a white porcelain surface will reflect just as much energy in a northward direction as will a brightly polished aluminum sheet.

But notice that Mother Nature again cooperates to make a well-designed solar collector work even better. In most areas, when you need the most heat there is usually snow on the ground. If your collector is installed at ground level, your entire yard becomes a kind of giant reflector, improving the over-all performance of your system.

By actual measurement, putting the system at ground level will improve performance by about 20% to 25% in the typical suburban back yard, and by as much as 30% in a very large back yard.

SNOW ON THE COLLECTOR

While we are talking about snow in the yard: One of the questions most often asked relates to snow build-up on those solar collectors which have a 60° tilt.

Obviously, snow falling at night or on a very cloudy day will stick to the covers and blanket the collector, preventing any collection from taking place until the snow melts. If you wish the maximum in collection, you should go out and remove the snow when the storm is over. If you are like me, however, and have inherited just the slightest tendency toward laziness, you can just wait.

Fortunately, because the covers tend to be very smooth and are aimed pretty much directly at the winter sun, by 10 or 11 o'-clock on a sunny day the snow will have slid to the bottom of the collector, leaving portions of it for sunlight collection. That heat, in turn, will melt the remaining snow

on the collector in a relatively short time.

Of course, if you purchase a 90° tilt (vertical) collector, snow build-up is not a problem.

TILT OF THE COLLECTOR

Another thing that you will hear a lot of varying opinions about is the tilt of the collector face. Years ago, someone guessed that it should be at an angle about equal to the latitude you live plus 15°. Since then, hundreds of "experts" have repeated that guess.

For the best radiation during the three midwinter months (which are, after all, the peak heating demand months) your collector should be installed at latitude plus 20°. For example, if you live in Columbus, Ohio, which is located at Latitude 40 degrees North, the collector should have a tilt of 60° from the horizontal.

That collector tilt controversy is more academic than anything else, however. Most manufacturers who mass-produce systems build the units with either a tilt of 60° or a tilt of 90° (vertical) for delivery everywhere in the United States and Canada. This allows the maximum economy in production.

The 90°-tilt (vertical) collectors do *not* deliver as much heat for each square foot of collector as do the 60°-tilt collectors. However, making the collector vertical permits tremendous economies in production, so that the solar furnaces can be sold for about half as much.

Since more than 99% of the mass-produced unitized solar furnaces come with either a 60° or a 90° collector tilt, the charts and tables in this book have been compiled with that in mind, rather than using the optimum tilt for each latitude. The Btu (British thermal unit) output figures used are, therefore, correct only for collectors having those tilts. It turns out that the difference, over the entire heating season, won't be a great deal, anyhow, since an other-than-optimum collector tilt simply means that collection will be at its greatest at another time of the year.

ORIENTATION OF THE COLLECTOR

If one is a real nit-picker, the collector shouldn't face due south, but about 10 degrees west of due south. Technically, if the collector faced due south, morning collection would equal afternoon collection (before-solar-noon versus after-solar-noon). But the outdoor temperatures in the morning typically are cooler than they are in the afternoon. Therefore, collector efficiency is usually slightly less in the morning. By orienting the collector about 10 degrees west of due south, you are heightening the afternoon collection to the detriment of the morning collection, but presumably collecting more over-all energy because you are doing more collecting when it is warmer outside.

As a matter of fact, it ain't all that critical. In the real world, do what looks best. If your house is lined up due north-and-south, you will probably want to install the collector parallel to a wall of the house. If your collector is properly insulated, you can install it on the southernmost wall of the house, staying within 30 degrees of due south, without noticeable difference in performance.

FINDING SOUTH

A question I get hundreds of times a year is this: "How do I find true south?" One can get very complicated with this. Some manufacturers publish maps of magnetic declination with their installation instructions, giving detailed instructions it takes an engineer to understand.

Like most other things, you needn't make it that difficult. One of the simplest ways to find south is to look at the survey map you got when you bought your house. North is indicated on that map. If you can't find that map, a city map will do the same job for you.

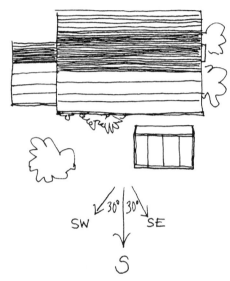

If you haven't got a reliable map, another simple way to find south is this: Look in your newspaper for today to find the exact time of sunrise and sunset. Halfway between those two is *solar noon*. (It probably won't be noon by the clock. Instead, it might be 1:16 P.M., for example.) At solar noon, go out in the yard where you plan to install the collector. Drive a stake into the ground, perpendicularly. The shadow will point north.

WATER SYSTEMS
VERSUS AIR SYSTEMS

Now comes some discussion of a point that has been a source of controversy for years in the solar-energy business. When you are building a new house, or adding solar heating to your existing home, should you buy a water system or an air system?

Remember Rule Number One: All collectors work about the same if they are insulated the same. Either collector system is fine, as far as performance is concerned. Either will do the same job. Unfortunately, they don't cost the same.

If a solar heating system were in every respect like a fueled heating system, a water/solar unit, using water to distribute heat collected from sunlight, would cost only about three times as much as an air/solar unit. For example, if you are building a new house and put in a natural-gas-fired forced-air system, it will cost you about $1,200. A gas-fired hot-water system with baseboard radiators will cost you about $3,600. This might lead you to think that a water/solar heating unit could be purchased for about $10,500 if, after shopping, you found that an air solar/system for your home would cost $3,500. Unfortunately, because you must store energy, and because of operating efficiencies at higher temperatures, a water/solar heating system must cost about four times as much to deliver the same output, measured in British thermal units, as a comparable air/solar heater.

The problem is due to the temperatures at which the solar furnaces must be operated. An all-water system—that is, one which uses baseboard radiators inside the home to deliver heat to the residence—must, of necessity, be a *high-temperature system.* An air system can be a *low-temperature system.*

That doesn't sound like an important difference, but it really is. Here's why: Let's take a unit operating in Chicago, Illinois. Because of additional heat loss at the higher temperatures, it takes 259 square feet of sun-radiation collector for a water system to match the Btu output, inside the house, of an air system powered by just 160 square feet of solar-heat collector. Not only will the water system cost more per square

foot of collector, just like the fueled heating systems, but also you will need nearly two-thirds more collector surface.

Although it will vary from manufacturer to manufacturer, you can expect to pay about four times as much for a high-temperature water/solar heating system as for a low-temperature air/solar heating system delivering the same Btu output. So, if it costs you about $3,200 for that 160-square-foot air system installed in Chicago, you will end up paying in the neighborhood of $12,800 for the high-temperature water system to do the same job. And that's an expensive neighborhood!

Well, if high temperatures are the problem, why not build a water system solar collector that operates at lower temperatures? Unfortunately, those baseboard radiators will not heat your house at lower temperatures. In fact, they have to be specially manufactured to work even at the temperatures of 160° (Fahrenheit) assumed for the above efficiencies. Most radiators nowadays are designed to operate at temperatures between 185° and 205°. Forced air, on the other hand, can keep the house warm with air temperatures that are only a few degrees higher than the room temperature.

One can, of course, build a *hybrid* solar heating system. That is, you can use a water-type solar-radiation collector and water storage, but then use an air/water heat exchanger to provide forced-air heat inside the house. Remember, though, that the collector will not work any better. And, as we shall see later, freezing can make the maintenance much more costly on the water-type collector. So you end up paying more initially and more in the long run, and you don't get a bit better performance from the collector. (You will possibly save a little energy on the pump circulating the collector fluid, but you'll lose that advantage in the energy consumption necessary to accomplish the air/water heat exchange.)

What do you do if you want to add solar heating to a house that already has baseboard radiators? I'm afraid the answer will not be really palatable. You should add ducting to your home and buy a forced-air system. This sounds horrendous, but it need not be. Many homes, particularly in the northeastern United States, have done just this, and the cost can be fairly reasonable if some good old common sense is used.

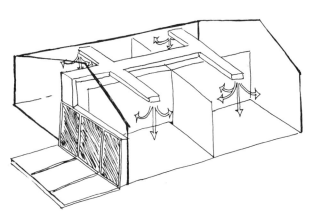

For example, in a single-story ranch-style house, air ducts can be added for as little as $300 to $500 by the simple expedient of running a central duct through the attic (or crawl space) and trunks from that duct to each room. Of course, an adequate return air duct must be provided, as well. In a two-story house, by finding closets on the first and the second floors that are directly in line, one can usually get an air duct from the basement to the attic and again use the central duct-and-trunk approach, although, of course, this will cost more. But even if it costs $1,500 or $2,000, it is much less expensive to do this than to go to a high-temperature water system.

RULE NUMBER 9

Unless you have money in great abundance, and an all-consuming desire to be parted from it foolishly, you will purchase a forced-air solar heating system for your home. This is true even if you are adding solar heat to a home that already has hot-water heat.

Let's face it! Solar heating ain't cheap. You will be paying from $1,500 to $5,000 for solar heating on your home. But to consider paying four times that much is plain, downright foolishness, unless you don't have to work for your money.

O.K. You've read many ads and many articles about water-type solar collectors. Why in the world are so many people building them? The reason is simple: Solar is an innovative technology. People are still imitating the mistakes of others. The literature is spotty. Many of the charts and tables that appear in this book are appearing for the very first time anywhere. There is a tremendous competitive spirit in the field; so manufacturers do not share their proprietary technology freely, but leave it to their competitors to learn the hard way. Also, if you have built a system that doesn't work very well but has already got tons of publicity, with your picture appearing alongside it, it is very understandable if you don't take out ads saying that it cost too much and didn't work worth a darn. One system built here in Colorado got that kind of publicity, then produced all of 1% of the home's heating needs during the first winter—and cost about $48,000. Needless to say, not much has been said about that performance by the engineers involved or by the city that sponsored it!

And us folks in the business haven't trusted your good sense. We've been afraid that, if we talked about these terrible failures, you would think that solar heating itself didn't work. If you got a new car that didn't work, you got a lemon; but if you got a solar furnace that didn't work, it wouldn't be a lemon, it would be a condemnation of solar technology.

And there have been plenty of problems in the growth of this new industry. Everything that you are being warned about in this book comes from experience. Some valiant pioneers have suffered through the problems and their solutions. Fortunately, for the most part those days are over.

But all these factors have contributed to a situation where reiterative mistakes have been made. Reports have gone to the government from researchers outside the government, and from scientists inside the government itself, that water-type solar-powered heating systems would never be practical for single-family residences. These reports, unfortunately, have been either suppressed or ignored, because to this day the government continues to spend your money and mine on stupid, incompetent projects in solar energy. And greedy corporations have been quite willing to take the taxpayers' money and build what the government wants: ungainly, overlarge water-type (and occasionally air-type) solar heating systems perched on rooftops around the country. The average person reading the newspapers has come to think that this is the way solar heating ought to be. It ain't so.

DEFINITIONS OF COLLECTOR EFFICIENCY

The single most deceptive practice taking place in the new solar industry has to do with something called "collector efficiency."

Prior to 1973, a definition of collector efficiency was used that was at least understandable. It compared the amount of energy *getting into* the collector with the amount *transferred* into storage, and ex-

pressed that ratio as a percentage. Using that definition of collector efficiency, the same collector would have the same efficiency regardless of the town in which it was used and regardless of the collector tilt.

Evidently that was too understandable, because a new definition was conjured up in 1974. The new definition compared the amount of energy falling on the *outside* of the collector covers to the amount of energy transferred to storage.

Doesn't sound like a big change, does it? Well, it makes it impossible to compare systems! Because, using that definition the very same collector would have a *different* efficiency in every city in the United States and Canada, and, furthermore, a different efficiency in the same town, for every different angle of collector tilt! Many manufacturers publish "collector efficiency" charts in their literature, using this new definition of efficiency. As if that information would help you make your buying decision! If manufacturers are going to tell you anything at all meaningful about their collectors, they should be using the earlier definition. But even that doesn't give enough information to allow you to evaluate the solar heating system.

What you also need is a storage-efficiency rating, because storage is an important part of the system, as well. Even that information almost requires that you be an engineer to interpret it. The manufacturers who are genuinely interested in allowing you to evaluate their product will tell you what the Btu output is, *inside the home, at the register*, after storage-heat losses. After all, what you are buying, presumably, is *heat*, not collector efficiency that is rated on a nonsense definition. The manufacturer should tell you how much heat you are going to get.

If I were buying a system today, I

would insist upon a nameplate Btu output rating from the manufacturer, one that was included in the manufacturer's warranty. If the manufacturer is ethical, he or she should be willing to tell you what the equipment will do—how it will perform—and be willing to stand behind that performance.

RULE NUMBER 10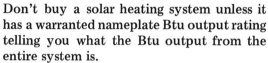

Don't buy a solar heating system unless it has a warranted nameplate Btu output rating telling you what the Btu output from the entire system is.

That rule seems so obvious that I am almost ashamed to include it. Surely you wouldn't buy a pig in a poke. Surely you would insist upon knowing what the Btu output would be, inside your home, before you purchased. But here is a scary fact: 99 out of 100 solar heating systems manufacturers do *not* tell you what the performance of their system is, in terms of Btu output!

If you don't have that vital bit of information, it is impossible to know how big the solar-energy-collecting unit should be for your home, or how much of the heating load it will handle.

Since so many manufacturers are recalcitrant about telling you how their equipment performs, you may simply have to outfox them. Since all collectors work nearly the same with the same amount of insulation, if you can determine how much insulation a collector has, and how much insulation the storage module has, you can tell what the Btu output of the system will be. There's no big mystery. In Chapter 13 of this book, do-it-yourself worksheets and tables will allow you to calculate—using simple multiplication and addition—the Btu output in your town, with the amount of sunshine in your town, month by month.

WHAT SIZE COLLECTOR?

As common sense will tell you, solar furnaces don't do the same job in Green Bay, Wisconsin, that they do in sunny Yuma, Arizona. It's both cloudier more often and colder, in Green Bay.

RULE NUMBER 11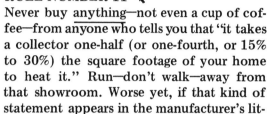

Never buy anything—not even a cup of coffee—from anyone who tells you that "it takes a collector one-half (or one-fourth, or 15% to 30%) the square footage of your home to heat it." Run—don't walk—away from that showroom. Worse yet, if that kind of statement appears in the manufacturer's literature, you're in bad company. Don't buy.

That salesperson, that company, and that manufacturer have to be confidence artists and rip-off specialists. There's no polite way to put it. Obviously it takes a different sized solar-radiation collector and storage unit for a given house in Yuma, Arizona, and that same house in Green Bay, Wisconsin. Only a fool or a crook would tell you otherwise.

What may not be so obvious to you, unless you are a heating contractor, is this: Heat loss is *not* proportional to the square footage of floor area in your home. For example, a properly insulated house with 1,000 square feet of floor area, with a crawl space or unheated basement, will be about a 4,500-Degree-Day house, if it is ranch style. (Meaning that it takes 4,500 Btu's to heat the house for 24 hours for each degree it is below 65° Fahrenheit outside. More about this in Chapter Ten.) A similar house with 2,000 square feet of floor area, or twice as much, will be only about a 6,800-Degree-Day house. That is only about half again as much heat loss. A 3,000-square-foot house, or one with *three* times the floor area, will

have *less than twice* the heat loss of the 1,000 square-foot house and will be about an 8,900-Degree-Day house.

So that idea of using floor area as some kind of "rule of thumb" doesn't even hold true in the same town, with the same builder, with different-sized houses!

I can't caution you enough, however, about that particular myth. It has gotten some real coverage. Even the staid National Bureau of Standards once published, in an early work on solar energy, "It is a good rule of thumb to use one-half the square footage of the home for the size of a solar collector." Fortunately, with scarcely a blush that statement was quietly removed from subsequent documents. You never get admissions of incompetency from government bureaucrats, but if you confront them with facts, you will get action, eventually!

It is important to notice, as well, that although collectors with the same amount of insulation work about the same, obviously a poorly insulated one doesn't work the same as a well-insulated one. In fact, a poor

collector may only work at 32% efficiency when it is 0° outside, while a good one is working at 72% efficiency.

RULE NUMBER 12
The only way you can size a solar furnace properly to your home is to perform a heat-loss calculation on your home and to know the useful Btu output of the solar unit in question.

SPECIAL CONSIDERATIONS IN WATER-TYPE COLLECTORS
If, after reading and understanding Rule Number 9, you still want a water-type collector, you're rich. And not conservative with your money. But you deserve advice, too!

RULE NUMBER 13
Do not, under any conditions, purchase a solar/water collector that has aluminum in contact with water. Demand copper.

Why? Because of a little gremlin called aluminum hydroxide. If the water or anti-freeze is running through aluminum tubing or passageways in your collector and if *any* of the rest of the plumbing in your entire system uses copper or brass fittings or tubing, the water must be "pacified." It must be distilled and de-ionized. No air must be permitted in the system. Fail on any of these points, and aluminum hydroxide forms. And that nasty stuff will eat right through the collector before you know it, turning it into a good imitation of a screen door.

It is, of course, possible to "pacify" a system in a white-lab setting. Out here in the real world it is simply silly. It's kind of like having the battery manufacturer tell you to use only distilled water in the battery, or the appliance manufacturer tell you to use distilled water in a steam iron. We get lazy once in a while and use tap water, right? The thing that makes me angry is that the collector manufacturers have known about the problem with aluminum collectors for years, and yet they still sell them to the unsuspecting public.

Finally, except for installations in the extreme Deep South, where absolutely no possibility of below-freezing temperatures exists, you must either use antifreeze or have a foolproof method for draining the collector at night and when temperatures of the collector get below freezing during the daytime.

I have yet to see a draining method that is foolproof, so be careful.

RULE NUMBER 14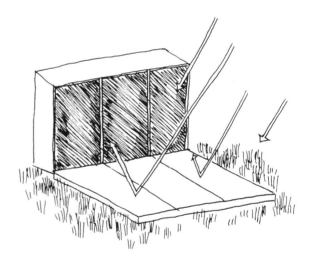
Be sure to find out how much antifreeze the system holds and how often it must be changed.

A poor engineering job can cost you a ton of money in this area. For example, one system has a 5,000-gallon storage tank underground, uninsulated, and not below the frost line. It uses 2,500 gallons of antifreeze. At a mere $5 per gallon, that's $12,500! If changed only every five years, that is $2,500 a year, average. Much more than the homeowner would have paid to heat the house with electricity!

It's bad enough that you and I get stuck as taxpayers for some of this kind of "solar engineering," but it would be instant bankruptcy if you got stuck with such a system on your home!

MIRRORS OR REFLECTORS
With a 60° collector tilt, and depending upon the latitude where you live, a mirror or reflector lying flat in front of the collector will

increase the amount of energy getting inside the collector by 25% to 30% in the midwinter months.

Now, it costs a heck of a lot less to put mirrors or reflectors on a collector than to make it 25% to 30% larger. It takes less energy to run the collector, too.

Another benefit of reflectors is that they boost the collector efficiency all day long, as well. Why? Because in a given collector at a given temperature, the heat loss is constant. If more energy gets into the collector as a result of adding reflectors, more energy is transferred to storage, making the *efficiency* of that collector higher.

That may be a bit confusing, so let's look at a concrete example: Despite the fact that manufacturers misleadingly publish collector-efficiency graphs showing performance of a collector under conditions of 350 Btu's to 450 Btu's per hour for every square foot—which is possible around noon on a sunny day—the *average* amount of energy getting inside such a collector with two covers is only about 130 Btu's per hour without reflectors and 155 Btu's per hour with reflectors, if you include the rotten collection times in the early morning and late afternoon. (*Your* solar collector works all day—not just at noon!) At 0° Fahrenheit outside, an R-30 insulated collector (review page 10) will be 67% efficient without reflectors, 72% efficient with reflectors. The heat lost through the insulation and covers will be 43 Btu's each hour, with or without reflectors, for each square foot of the collector. For the collector without reflectors, 87 of the 130 Btu's are transferred to storage; 112 of the 155 Btu's of the collector with reflectors are transferred. Thus, not only does more energy get *inside* the collector with reflectors, but also a larger percentage of the energy is *transferred* to storage.

The mirrors or reflectors should be changed, month by month, to obtain the optimum amount of reflection into the collectors. That's theory, and it's true. However, out here in the real world, again, who needs all that bother? The peak demand for heating comes during the midwinter months —December, January and February. It turns out that the optimum mirror angle for those months is very nearly zero degrees (flat on the ground). Most people (myself included) are lazy enough to be content with leaving the mirrors flat all winter long.

What should the reflector be made of? Shiny aluminum, dull-finish aluminum, silvered glass or white porcelain? The answer is that it really doesn't matter. They all work about the same, as far as getting the energy inside the collector is concerned.

That may seem a bit contrary to common sense at first, since it would seem that the shiny aluminum or silvered glass would be best. If we were talking about the kind of reflection where the light is reflected in neat parallel rows, that would be true. Remember that earlier we talked about the fact that ordinary clean snow reflected 87% of the energy striking it in a northward direction?

Feel the mirror or reflector surface when the sun is shining on it. It is not very warm. Why? Because, depending on the angle at which it strikes, less than 10% of the energy striking the surface is absorbed. What happens to the rest? It is reflected in a northward direction. Obviously, if a white porcelain or dull aluminum reflector is used, it is primarily diffuse radiation that is reflected. The collector doesn't care!

Does that explanation satisfy your common sense? Remember that, at an extremely oblique angle, even black paint reflects 33% of the energy striking it. At the angles at which your collector reflectors will be operating, about 90% of the energy striking them will be scattered toward the collector. Because of the angles at which that light arrives at the collector covers, a much larger percentage is reflected away by each air/glass interface; but from 25% to 30% additional energy gets inside the collector.

RULE NUMBER 15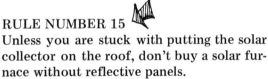
Unless you are stuck with putting the solar collector on the roof, don't buy a solar furnace without reflective panels.

If the manufacturer has your interests at heart, reflectors will be provided as an integral part of the unit simply because this increases the Btu's you get inside the house for each buck you spend.

The so-called "A-frame" or "backyard" solar furnaces have a further, patented feature: The reflectors are hinged like doors and can fold up over the collector during the summer, to give additional protection to the collectors during high yard-use times, to increase the useful yard area in the summer, and to stop collection when it isn't needed.

Be sure to cast a skeptical eye at the construction of the reflective panels, however. How strong are they? Are tie-downs provided, so that they won't blow away in the first windstorm? If they are flimsy, has the dealer provided for a supporting framework as part of the cost of installation? Are they so flimsy that you will have to install a fence to keep the kids and the dogs off them, for fear of damage?

COLLECTOR MATERIALS
Take a good look at the construction of the solar collector you plan to buy. As we saw earlier, it is important to insist upon seeing the results of a seven-day stagnation or stasis test.

But there are dozens of plans for solar collectors being sold nowadays that give abysmally misleading recommendations. Remember that the properly insulated collector will achieve temperatures approaching 400° Fahrenheit. That means that neither polystyrene foam nor bead board can ever be used in a collector or a storage unit because they both melt at 155°! It's amazing how "expert" some of these folks are at telling you how to do something that they obviously never have done themselves, or they wouldn't make such a stupid recommendation.

Even polyurethane foams melt at 300°;

so they cannot be exposed directly to the heat inside the collector although they are fine for storage. For safety reasons, however, if polyurethane foam is used it must be faced on all sides by some other material.

If asbestos is used for the insulation, be sure that it is totally sealed away from any air that could possibly get inside the house as it's a dangerous health hazard.

One of the quickest ways to see if all these considerations have been met by the manufacturer is to ask if the solar unit in question conforms, *as a system*, to the building codes.

2 Other Types of Solar Collectors

CONCENTRATING COLLECTORS

Technically, the flat-plate collectors that use mirrors or reflectors are concentrating collectors, having a 1.3 concentration factor. However, for the purposes of this discussion we will only consider collectors with a higher concentrating capability as "concentrating collectors."

RULE NUMBER 16

Do not buy a concentrating collector solely to heat your home. The only justification for using a concentrating collector is to obtain higher temperatures.

You have probably used a concentrating collector: an ordinary magnifying glass. You can concentrate the energy falling on the entire glass to one small point and achieve temperatures high enough to start a fire.

Because of the higher temperatures involved, it is easy to make the mistake of thinking that concentrating collectors are more efficient or that they collect more energy than a low-temperature flat-plate collector.

Nothing could be further from the truth. The higher the temperature, the greater the heat loss. The more heat loss there is, the less the efficiency.

No more energy falls on a square foot of concentrating collector than falls on a square foot of flat-plate collector. The only difference is the temperature achieved.

A concentrating collector has a number of disadvantages: It must be designed to track the sun, it requires direct sunlight in most cases, although some have been designed to make use of diffuse energy, as well, and it must be kept scrupulously clean.

For solar-powered cooling applications, the concentrating collector may hold some promise, simply because the absorption chillers and Rankine cycle coolers require temperatures in the 200° to 250° (Fahrenheit) range to function efficiently, and flat-plate collectors have rotten efficiencies at those temperatures.

Work is being done, as well, on using concentrators to focus more energy on photovoltaic cells in an attempt to bring down the cost for that kind of electrical genera-tion. Concentrators can also be used to cre-ate steam to power turbines for the genera-tion of electricity. Both these applications are terribly expensive, however, and I wouldn't expect to see them used on family residences. At present, it would cost at least $60,000, retail, to provide a modest 18 kilo-watt hours per day for your average electri-cal consumption for TV and appliances.

There are concentrating collectors on the market. One of them probably should get an award for the rottenest engineering ever done, in that it is a tracking high-tem-perature collector with just 1/2 inch of fi-brous glass insulation on its sides! Others claim efficiency improvements of 40% to 70% due to their tracking of the sun. Untrue. On the shortest day of the year, December 21, at Latitude 40° degrees North, the im-provement would be on the order of 13% in terms of the radiation transmitted to the inside of the concentrating collector, and losses due to higher temperatures would probably cancel out most of that advantage.

EVACUATED-TUBE COLLECTORS
Another high-technology collector is the evacuated tube. Again, higher temperatures are produced. Because of a vacuum sur-rounding the absorber, the losses at higher temperatures are significantly reduced. But, as we said before, higher temperatures pro-duce inefficiency in the most important part of a solar furnace: the storage system. At 85° Fahrenheit they are scarcely more ef-ficient than a properly insulated flat-plate collector.

As can be seen from Figure 5, the evacu-ated tube is very much like a fat fluorescent light tube or a radio vacuum tube. It does represent high technology being applied to modern-day problems. Unfortunately, it is

unlikely that better efficiency results per unit of gross area of collector, and the cost is much, much higher.

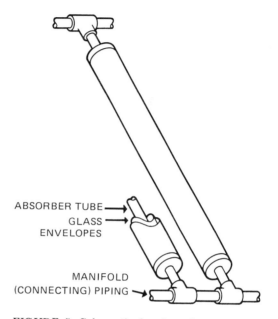

ABSORBER TUBE
GLASS
ENVELOPES

MANIFOLD
(CONNECTING) PIPING

FIGURE 5: Schematic drawing of evacuated-tube collector.

The evacuated tube suffers from one fatal flaw, in my opinion. One that will probably prevent it from ever being used widely for heating applications. Because of the vacuum inside, the tube will present an irresistible attraction to every red-blooded boy and girl on the block. Once the word gets around. Because when it breaks, the evacuated tube sounds like a cherry-bomb firecracker. BOOM! If the tubes happen to be installed too close to one another, the implosion of one may detonate the one next to it. I'm not sure I could resist that chain reaction myself, after one too many drinks!

COMPOUND PARABOLIC COLLECTORS
Another collector type is the so-called "crab eye," which was first invented in Japan by

a group of governmental scientists about 10 years ago. It was "re-invented" at the University of Chicago in about 1973. A form of this collector is shown in Figure 6.

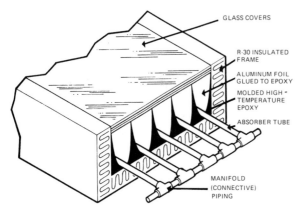

GLASS COVERS

R-30 INSULATED FRAME

ALUMINUM FOIL GLUED TO EPOXY

MOLDED HIGH - TEMPERATURE EPOXY

ABSORBER TUBE

MANIFOLD (CONNECTIVE) PIPING

FIGURE 6: Compound parabolic, or "crab-eye," collector.

There are other varieties of concentrating collectors. They all have a great appeal, simply because the temperatures they generate are impressive and they look more efficient, more "technological." Ironically, most concentrating collectors that are available today have even less insulation than the low-temperature collectors, despite operating at higher temperatures and requiring more, not less, insulation.

Such ill-contrived design will probably cease when you, the consumer, begin asking questions that show the manufacturer that the product won't sell without proper, energy-efficient engineering.

RULE NUMBER 17

Do not purchase a concentrating collector for home heating or cooling without seeing a warranted Btu Output Rating for the entire system, inside the house, under real-world conditions. Compare that rating with the rating for flat-plate collectors delivering the same output and compare total costs.

3 Storage of Thermal Energy

THE IMPORTANCE OF STORAGE

O.K. So now you know what to look for in solar collectors, right?

From the attention given to solar collectors, most people have the mistaken impression that the solar collector is the most important part of a solar furnace.

That is simply incorrect. Of course, you have to have a collector. Obviously, the unit isn't going to work without one. But, as we will find out later, you rarely need heat in the daytime, even in midwinter, if your house is properly insulated and the drapes are open.

So a collector on a home, without an accompanying, properly insulated storage unit, is nearly worthless. Because of the energy required to run a pump or a fan and the low heating demand in the daytime, it could cost more to run than it would save in energy.

There are companies selling collectors for daytime-only use on homes. It's a *cheat.* There are applications where daytime-only heating makes sense: for example, on big shopping centers or big factories where there

are few windows and where the buildings are used only in the daytime. But it doesn't make sense to provide collectors without storage for use on your home.

The key component of any solar furnace is the storage module. Even though it is nothing more than a well-insulated box of rocks or a well-insulated tank of water.

Unfortunately, dozens of myths and mistaken theories have sprung up around storage, and how it ought to be designed. This chapter, therefore, is very important to your understanding of solar heating devices and to your evaluation of competing systems.

HOW LONG WILL IT HOLD HEAT?

One of the most deceptive practices in the solar-energy industry is to tell you how many days or weeks a solar storage unit will hold heat.

For example, a properly insulated storage unit (water or rock) located out of doors in ambient temperatures averaging 30° Fahrenheit, starting at a mean temperature of 120°, would still have useful temperatures (above 75°) 12 full days later if the heat were not to be used for heating the home.

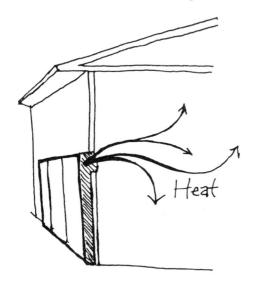

Of course, the tricky words there are, "if the heat were not to be used for heating the home." Such a statement is really talking about the insulation and the heat-holding design of the storage unit, and, indeed, it is good to know that the storage system has been designed to keep heat-loss to an absolute minimum. This is important because the collector has to be oversized to make up for losses from storage, and that gets you in the hip pocket.

But let's use our common sense again. If a solar furnace is sized to give you 90% of your home-heating requirements during the 273-day heating season, it will only supply about 75% of your heating needs in December and January. What does that mean? It means that the solar furnace is supplying, *on the average*, 75% of the heat needed each day during those two months. If the sunshine striking the collector were exactly equal every day, you would only need to store heat for a maximum of three-quarters of one day. Of course, some days are sunny

and some are cloudy, so that some days you are collecting enough for as much as a day-and-a-half of heating; so the storage unit should hold heat without major losses for that day-and-a-half.

Further, a system that supplies 75% of your heating needs in January will supply 100% of your heating needs in September, so that even on a cloudy day in September the collector will probably collect enough energy to heat the house that night. So, again, if the system holds heat without big losses for a day-and-a-half, that is an adequate time period.

RULE NUMBER 18
Don't be taken in by talk about how long the storage unit will store heat, particularly if the time periods talked about are more than a day-and-a-half.

HOW MUCH STORAGE IS NECESSARY?
In the early days of solar development, those who designed equipment thought that *huge* amounts of storage were necessary. There were many articles and learned treatises written about various storage materials.

It was an understandable mistake: The assumption was that the problem would be in designing *enough* storage capacity. If that had really been the problem, then one would have to worry about how much heat a pound of various materials would hold. Because the space that storage took up could become very important.

And if that were true, without question water would be the preferred material to use for storage because water holds more thermal energy per pound than other materials do. For example, one pound of water will hold five times as much heat energy for each degree of temperature rise as one pound of rock. That is a characteristic of these ma-

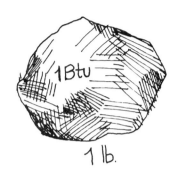

terials. Early solar designers, though, began to make a big thing out of that difference. (Even though saying such a thing was about like saying that a 55-gallon drum holds 11 times as much gasoline as a 5-gallon drum. It is true. It is also obvious.)

Actually, storage doesn't have to be any bigger than needed to store about a day-and-a-half's worth of collected energy, if that much. In fact a good rule of thumb is to provide enough storage so that one full sunny day's collection in January will create a 75° increase in temperature in storage. If you did that for a solar furnace in Denver, you would need to provide a cubical storage unit about 4-1/2 feet on a side, if you used rock storage, while a water storage tank would have to be about 3 feet in diameter and 4-1/8 feet high to do the same job. Make a big difference? Heck, no.

It's kind of like this: If you are out of gas between two towns and you have to walk into town 3 miles and carry back enough gas to drive back to the station, which would you carry: a 1-gallon can or a 55-gallon drum? By the reasoning of some of the early solar theorists, you would carry the 55-gallon drum—not because you *needed* more, but because the 55-gallon drum *held* more!

HOW WELL DOES IT STORE HEAT?

This is really the more pertinent question to ask about a storage system: How well does it store heat and what are the heat-transfer characteristics?

The more efficient the storage unit the smaller the collector needs to be and the less you will have to shell out to keep your family warm.

It turns out that even though water may hold *more* heat, some solids hold heat much *better*, even in containers with the same amount of insulation. Why? Because the water circulates, tending to keep the hottest temperature it contains evened out and contacting the entire surface—top, bottom, sides—of the tank containing it. It makes an excellent thermal bond with the tank and loses heat through the insulation at the fastest possible rate.

If you fill that same tank with pebbles instead of with water, it will not lose heat as quickly. Here's why: After a period of time, the pebbles nearest the top, sides and bottom of the tank will be cooler than the pebbles in the center of the tank because of heat loss through the insulation. If the pebbles are small enough (about 1-1/2 inches in diameter or less) the air spaces between the pebbles will be small enough so that the air surrounding each pebble will not convect (circulate) naturally. This "dead" air acts like a wrapping of insulation around each pebble, so that the heat contained must be transferred from pebble to pebble only through the small points where the pebbles touch each other, by conduction (contact heat-transfer). Notice also that the pebbles will only contact the walls of the tank at points, again slowing the rate at which the heat will be transferred.

Our research facility was probably the first in the country to test storage materials against each other with the question in mind, "How *well* will it store energy?" It turns out that nothing loses heat faster than water. Rocks, ceramics and glass all test about the same. But rock is the least expensive and is readily available all over the country.

And I have never seen a rock wear out, rust, freeze or leak.

This is all fairly academic, however, since a properly insulated water storage unit located outdoors and above ground is about 94% efficient at an outside temperature of 30° Fahrenheit. (For conditions, see notes to Tables 16-36, Percentage Efficiency of Various Storage Modes, Chapter Thirteen.)

INSULATION OF STORAGE

As you may have guessed already, the really important part of storage is the amount of insulation it has.

Just as with the solar collector, the more insulation there is on the storage, the better it will serve its function.

RULE NUMBER 19

Don't buy a solar furnace where the storage unit has less insulation than you have in your attic, if it is to be located outside or below ground (R-30 minimum). If the stor- **age unit is to be located inside the house, don't accept less than R-15 insulation.**

Again, it just stands to reason. If your solar furnace is working at all, the temperatures in storage will be higher than the temperature you keep your home. Therefore, if the storage is outside your house it should be as well insulated as your house, at least. A storage module inside your house should be well enough insulated to keep the heat in storage until your house thermostat calls for heat. Many solar-technology engineers will say that the storage unit doesn't need that much insulation if it is inside the house because the heat lost goes to the house, anyhow. Well, that's daffy! In September and October, the solar heating system may produce as much as 1-1/2 times to 6 times the heat you need to keep the house warm. If that heat is transferred to storage and then, because of rotten insulation, that heat is lost to the house, you may see temperatures *as high as 100° Fahrenheit in the house!*

The point of storage is to *store*, not to *lose* heat.

Underground storage is fraught with many potential problems in insulating. Even if the unit is properly insulated, it must also be perfectly waterproofed because, with most insulation materials, if they get wet their insulation value is lost. I am of the opinion that it is senseless to put storage underground. Why? Because, even if you are skillful enough to seal the unit *perfectly*, the cost of doing that is more than simply building the same unit above ground.

I might pass on another example of what not to do, while we are talking about underground storage. A system was built here in Colorado using a 5,000-gallon tank of water for storage, with the tank being installed in the ground under the garage. The engineer in charge did not put *any* insulation on the tank. When I asked how come it was being done that way, he patiently explained to me that any heat loss from the tank would heat up the ground around the tank, and then, when heat was used in the house, the ground around the tank would be a bit warmer than the tank, and heat would flow into the tank. I allowed as how he would have to heat up the entire world before that would work, but he suggested that, as an engineer, he knew what he was doing. Well, even below the frost line, down where it doesn't freeze hard, the ground temperature in that city drops to 37° in January; and this tank wasn't below the frost line. What happened was that the tank cooled off to the ground temperature and stayed very close to that temperature, despite the valiant efforts of some 750 square feet of collector working to warm it! (750 square feet of collector is enough collector for *five* properly insulated houses in that particular city!)

"CLOGGING" OF PEBBLE-BED STORAGE

One of the most ludicrous myths to make the rounds concerns pebble-bed storage. The typical pebble used is from 3/4 inch to 1-1/2 inch in diameter. If you look at a stack of such pebbles, it is a bit hard to imagine how the air makes its way through that dense mixture. Probably that appearance of impermeability gave rise to the myth that, after a year or two of use, pebble-bed storage would have to be cleaned out, because it would have settled and gotten clogged with dirt. Some versions have it that the rock crumbles a bit from expansion and contraction.

Don't worry at all about this particular myth. Here's why: Solid rock weighs 160 pounds for each cubic foot. For a rock storage unit to get clogged, you would have

to add about 60 pounds of dirt to the system for each cubic foot, or, in a typical storage unit, 15,600 pounds (7.8 tons) of dirt! That is a dump-truck load of dirt.

EUTECTIC SALT STORAGE

There is something inherent in mankind that inspires a love of complicated things. It must be this love that gives rise to the complicated high-technology storage system called "eutectic salt storage," or "phase-change storage." A special salt is used, typically one called "Glauber's salt," which melts at a fairly low temperature. Each pound of this salt, when melted (having changed phase from solid to liquid), stores 108 Btu's. About 2,000 pounds are all that is required for most residential applications.

The major problem with phase-change storage lies in the fact that the process requires that a great surface area be exposed for each pound of salt. The accepted formula is 25 times as much surface area as volume contained. Thus, if you have one cubic foot of salt, you need 25 square feet of surface area in order to accomplish the phase change.

There is absolutely no advantage to eutectic salt storage over ordinary water or rock storage, in terms of performance. It doesn't work any better, but it does give the salesperson a lot to talk to you about and it certainly gives the appearance of modern high technology at work. Unfortunately, not only does it cost a bunch more when new, but it must be maintained, as well. After about 400 phase changes, the material breaks down and must be replaced. Corrosion is reported to be a major problem, too.

Presumably, the only reason you would spend more money for a given storage unit is that it works better. And if that is the criterion you use for spending your money, you will not buy eutectic salt storage. It's that simple.

Unless you are rich and foolish, or are spending the long-suffering taxpayer's money.

HIGH-TEMPERATURE VERSUS LOW-TEMPERATURE OPERATION

In our discussion of solar collectors, we saw that at higher temperatures the performance of flat-plate collectors fell off considerably. Naturally the same problem besets high-temperature storage. The efficiency of the very same storage unit operating at high temperatures, compared with one operating at low temperatures, falls off from 82% to 61% at 30° F outside. (See notes to "Storage" in Chapter Thirteen.)

Probably that kind of decline in performance, by itself, could be lived with. But operating the collector at high temperatures at the same outside temperature results in an efficiency drop from 81% to 55%. That means that the total over-all system efficiency has dropped from 66% to 34%. So, again, it simply doesn't make sense to operate a system at the temperatures that would be necessary to provide hot-water baseboard heat.

THERMAL LAG

The fatal flaw in high-temperature systems is something called "thermal lag."

Suppose that, because of weather conditions, no collection takes place for three days and the storage temperature has dropped to 152° from 160°—the storage being a 1,500-gallon water tank, insulated R-15 and situated in the basement. Just to get that tank of water heated back up to 160°, which is the minimum temperature where the heat would be useful, you would need 110 feet of collector operating all of one average day in Sacramento, California. By comparison, a properly designed low-temperature system with only 90 square feet of collector would provide 90% of the heat required by a 1,000 square foot (5,000 Degree-Day) home in Sacramento. So, in order to operate a high-temperature system, not only would you have to over-size the collector because of poorer efficiency, but you would also have to add 110 feet of collector just to take care of thermal lag. That's between $2,200 and $6,600 worth of extra solar-utilization equipment!

STRATIFICATION

There is one advantage to pebble-bed storage units over other kinds. It lies in something called "stratification." Unlike water, for which you know that when one part of the storage tank is at a given temperature, the rest of the tank is very nearly at that same temperature, pebble-bed storage has extreme temperature differences from one side of the storage unit to the other. For example: Typically, the side of storage nearest the inlet duct to the collector will be at about 60° F, while the side nearest the collector outlet duct will be at about 110°. They are called the "cold" side and the "hot" side of storage, respectively.

If the system is properly engineered, the top of the storage unit will *not* be hotter than the bottom. Why? Because if that were true, natural convection (circulation of air) would be taking place in storage, which would increase heat loss.

Here is where the benefits of stratification make themselves known. When heat is removed from storage to heat the house, the air flow through storage is in the opposite direction from the air flow from the collector, which heated it in the first place. In other words, the return air from the house enters storage on the "cold" side and is removed from storage and directed to the house from the "hot" side of storage. In actual practice, this means that as much as 90% of the storage's pebble bed can drop below useful heating temperatures (typically about 75°) and there will still be heat energy stored in the other 10% of the pebble bed, on the "hot" side. When even the last 10% of the pebble bed is down to 75°, the system turns off, the fueled furnace in the house takes over, and everything waits until collection can again take place.

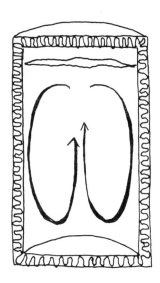

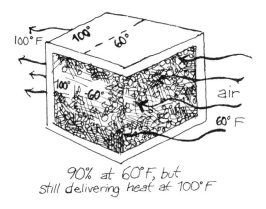

90% at 60°F, but still delivering heat at 100°F

Notice that this is quite different from water storage. There, the instant that *any* part of the water is below useful temperatures, the *entire* tank is below useful temperatures and the system turns off, with the fueled furnace taking over.

More important, because of this stratification there is very little thermal lag. With the average temperature of storage at only about 65° after the system drops below useful heat-delivery temperatures (called the "downpoint"), the heat losses from storage are very small. When collection again resumes, the "hot" side of storage gets heated first. Once about 10% of the pebbles are heated the system is back on stream, ready to supply useful heat to the home. This is in direct contrast to the water heat-storage unit, which must *all* be heated above the

downpoint before useful heat delivery to the residence can take place.

RULE NUMBER 20
If you have any choice, insist upon pebble-bed storage.

That rule may be a bit controversial but is included here because of the minimization of thermal lag, which means that you, as a consumer, will get more unit for less money, more Btu's for every buck.

By the way, when purchasing a system with pebble-bed storage, look very carefully at the directions the air flows during collection and during distribution of heat. In order to benefit from stratification, there must be *counter-flow*; that is, the collecting air must flow through the storage in the opposite direction from the air flow that will heat the home.

LOCATION OF STORAGE MODULE
As we saw earlier, it is less troublesome to put storage above ground than below ground, if it is located outside the house.

space saved with storage outside

But should you put the storage inside the house, instead of outside? With identical amounts of insulation, the over-all performance of storage located inside the home will be slightly better than that located out-side. However, the difference is not large: Only about 4% difference, with the same insulation.

Even if installed inside the house, storage should have very good insulation. So the cost difference for the container is negligible. It will cost a lot more money in the long run, though, if you put the storage inside your house, even though the original cost of the storage container is the same: Nowadays it costs anywhere from $12 to $20 per square foot of basement area in building costs. Even if only a small area, 8 feet by 10 feet, is taken up by the storage container, you have lost 80 square feet of usable basement area. That is between $960 and $1,600 in construction cost. You save all that money by locating the storage out of doors, simply because you gain the use of that 80 square feet of space.

4 Delivery Systems

FANS AND PUMPS

Fans and pumps have been around for a long time. It is quite easy to overlook their importance and their impact on the over-all efficiency of the solar heating system.

If the system is an air unit, the collector fan and distributor fan should not be larger than 1/2 horsepower and should not be smaller than 1/6 horsepower. If it is a water system, the pump motors should not be larger than 1/10 horsepower nor smaller than 1/30 horsepower. (The latter can be smaller because water is a better heat-transfer medium.)

The fans or pumps should be easily accessible, for service and maintenance, but should be protected from easy (and dangerous) access by children.

LOCATION OF FANS

In air systems, the fans and motors should be located inside the circulatory system. That way, any heat lost from the motors is still retained *inside* the system and is either stored in the pebble bed or directed to the house. Therefore, the electrical energy you pay for to run the fan or fans ends up, for the most part, as heat inside the home.

ENERGY CONSUMPTION OF MOTORS

The energy consumption of a water/solar heating system is typically only a quarter to a fifth that of an air system. Therefore, one would be justified in thinking that the water system would be much better in the long run. Indeed, the monthly cost of operation is less with the water system, if figured only in terms of electricity used.

Let's look at a real-world example. In Greensboro, North Carolina, it would cost an average of $3.24 per month to run the fans on a well-designed air system, and only about 65 cents per month to run the pumps on the water system, at a rate of 4 cents per kilowatt hour. That's about $23.33 less per heating season for the water unit.

Unfortunately, the water system has a cost associated with it in another area that more than makes up for the saving in the cost of electricity. Even with heat exchangers, it is not unusual to have 25 gallons of antifreeze in the collection circuit. If that is replaced every three years as part of the maintenance program, at a cost of $5 per gallon, the average annual cost for that maintenance is $41.67, or nearly double the savings on electricity.

It turns out that, if one engineers a low-temperature water system, one needs to go to an air-water heat exchange, and then one part of the water system is still using a fan; so the electrical savings diminish.

Of course, as electrical costs increase the savings will grow, as well. But the cost of antifreeze can be expected to increase at about the same rate, too.

EQUIPMENT

A direct-drive motor-and-fan combination is generally not considered as good as a belt-driven motor-fan combination. The belt-driven system costs a bit more initially. Both have very good longevity, however, and I suspect that the differences in the real world are not that great.

If you are using a water system, you should insist upon stainless steel pumps. They don't cost that much more.

Look for permanently sealed, permanently lubricated ball bearings on motors, pumps and fans. Other bearings, which need lubrication on a regular basis, are not as good and require you to remember to do the lubrication. You're likely to forget, and so cost yourself a bundle. It really doesn't cost much more (on the order of $2) to get sealed ball bearings.

5 Solar Furnace Controls

PROPORTIONAL CONTROLLERS

Today, nearly all solar heating systems have something called the "proportional controller." As the pioneer of this type of control, I am extremely satisfied to see an early judgement borne out by others in the industry.

The proportional controller has at least two temperature sensors, one in your collector and the other in your storage. Some units have two or more sensors in the storage, particularly if it is fine-pebble storage, simply so that an average storage temperature can be determined. The operation of that portion of the controller is extremely simple: Any time that the collector gets to be 10°F hotter than the storage, the collector fan or pump is turned on. It continues to run until the difference is no more than 5°. Then the fan or pump turns off.

The other part of the controller is tied into your home's thermostat and to the gas valve on your fueled house furnace. Additionally, this part of the controller reads the temperature of the storage. Any time when your thermostat calls for heat, the so-

lar furnace is called upon to heat your home. If the storage is above the downpoint temperature the solar furnace can supply heat and your gas furnace never turns on. As the storage is depleted, however, and nears the end of its useful heat, it may not be able to keep up with the total needs of the house. In that case, *both* the solar furnace and your house furnace run at the same time. Of course, once the storage temperature is below useful temperatures (typically 75° in air systems), the solar furnace is turned off and the fueled furnace does all the heating.

All this should be totally automatic. You should be simply setting your thermostat just the way you always have.

Unfortunately, some systems are still being marketed that supply a simple thermostatic control on the collector: Whenever the collector temperature rises above a certain preset temperature, the collector fan or pump turns on. This is an extremely unsound control system, since it is very possible to have the storage heating the collector at the end of the day and dissipating heat through the collector until all the storage has cooled to the preset temperature on the thermostat.

RULE NUMBER 21

If the solar heating system does not have a proportional controller—one that compares temperatures in storage and on the collector to determine when to turn on the collector pump or fan—don't buy the system.

Engineers who design solar heating equipment often make the mistake of thinking that, because solar energy is free, they don't have to worry about conservation of that energy. Unfortunately, the utilization of solar energy is *not* free, because the equipment to harvest it costs money—and a lot of it. The better the equipment is designed to conserve energy, the smaller the equipment can be and the less it will cost you. You must set the cost of the equipment against the "free" solar energy you get out of the system.

6 Solar Devices for Heating Domestic Water

THE COLLECTOR

If you are going to add solar-heated domestic water to your home, be very careful.

RULE NUMBER 22

Do not buy a solar hot-water heater with a water-type collector unless you have seen a certificate of compliance with all building codes.

This is such a problem that my recommendation is that you only buy a system that has an air-type collector, unless you are fortunate enough to live far enough south so that freezing isn't possible. When the collector has antifreeze in it, it must go through a heat exchanger in any case, just as does the air-type collector; but, in order to be safe, it must be a relatively inefficient double-wall type.

Why? If a leak develops, you do not want poisonous antifreeze in your drinking water.

Unfortunately, many units have been sold without this vital safeguard. It is really simpler, and less costly to you, to buy a solar air-type system for heating water.

Most of the consumer abuses in the infant solar-energy industry are occurring in the domestic-hot-water area. There are thousands of really rotten systems being sold for this purpose. To give you an idea of just how rotten, and just how widespread the incompetent design is, let me give you an example. A New England utility company recently put out 100 different systems for heating domestic water in homes in their service area. The *best* of the systems they chose provided only 37% of the hot-water needs, and the worst was so abysmal that the utility refused to report the results. The *average* performance was a ludicrous 17%! This fantastic performance came from systems that would cost you, the consumer, as much as $2,500, installed! Yet there was absolutely no need for such poor performance. You guessed it. Most of the systems tested had *less* than 3 inches of fiber glass on the collector, some only had one cover, etc.—violations of the simple rules of common sense we have already covered.

To get excellent performance—90% of the energy supplied by solar equipment—a family of two living in New York City could install a properly designed system having only 34 square feet of collector and costing about $1,000. Still not cheap. But it is ironic that rotten systems cost $2,500 while decent ones cost about half as much. I believe I have seen more of the "get rich quick" mentality in the hot-water-producing part of the solar-power industry than in the heating part.

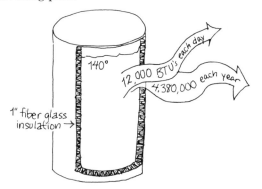

One of the big problems with your hot water tank right now is poor insulation. If it is typical, it has 1 inch of glass-fiber batt wrapped lackadaisically around the tank. You have to buy 12,000 Btu's from your utility *every day* (with a 50-gallon tank) just to make up the heat lost because of poor insulation. If your family size is two members, that represents fully 30% of your monthly bill. For four family members, it still represents 20%. Simply adding R-30 insulation to the hot water tank in your home can reduce that 12,000 Btu heat loss to a mere 1,350 Btu's each day.

Another problem with your present hot water tank, if it is gas-fired, is the design. I wonder if the gas companies designed them especially to use up gas! Why? Because the vent to the outside runs right through the middle of the water tank. You couldn't

make it lose more heat by natural convection (with the draft up the chimney) if you tried. Unfortunately, the present-day market doesn't seem to have any gas-fired units without this feature; the alternative is to buy an electric water heater. But the electricity costs about three times as much as the gas. So that isn't a good idea, either, since the heat lost up the chimney isn't enough to warrant the changeover. Yet.

One of the things most manufacturers neglect to tell you is that the size of your family is very important to the size of the collector you should put on your home. For example, we saw above that a two-member family in New York City would require 34 square feet of solar collector (approximately) to supply 90% of their heating needs. A four-member family would require 60 square feet for the same percentage, and a six-member family would need about 84 square feet. Remember, these numbers are correct only for a properly designed system: R-30 insulated double-cover 140° F collector with horizontal reflectors and R-30-insulated water tank. And the numbers would be different in different cities.

On page 88 you will find a Solar-Collection-Gram for Hot Water Heaters that will simplify the determination of how much of a unit you need in your home town (Nomogram 2).

It will cost you a lot less, generally, if you add solar hot-water heating at the same time that you add solar space heating to your home. But be certain that the two units have separate controls, or separate control sections, since you do not want to run the space-heating solar collector all summer. Because you are able to use the solar collector for hot-water heating all summer, however, your pay-back time for the solar water heater is about 20% shorter than for the house-heating equipment.

SECTION TWO

Making Your Home an Energy-Efficient Passive Solar Heating System

7 The Most Important Part of a Solar Furnace

YOUR HOME

Your home is a better solar furnace than you think. For example, in the month of January in Concordia, Kansas, a three-bedroom house with 160 square feet of windows equally distributed on the four sides of the house will have 50,000 Btu's per day transmitted through those windows to the inside of the house, on the average, allowing for local cloudiness. That's 1,500,000 Btu's per month that will get into the house if the drapes are open. If you are heating your home with electricity and are paying just 4 cents per kilowatt hour, that works out to saving $17.50 per month!

Obviously, the amount of energy that gets into a house varies from city to city, but it is always surprisingly large.

Simply opening all the drapes in your house in the daytime and remembering to close them at dusk will turn your house into a solar furnace, of sorts. That energy that comes through the windows is energy you don't have to make up with fuel.

Those windows can be two-faced, though: If they are not covered at night, they can lose more heat for you than they gained in the daytime. The kind of drapes you have can make a big difference, too: They should be heavily lined to be better insulation.

To see just what a good solar collector a house is, look at the house across the street in the daytime. Look at a window where the drapes are open. Looks black, doesn't it? Despite the fact that the room behind it has light-colored walls, more than likely. Why? Because the room is large, and the light that gets inside is bounced around so often that most of it is absorbed and very little is reflected back out of the room. The light that is absorbed becomes heat to reduce your neighbor's heating bill.

Recently I saw a new invention being heralded, which used black-painted venetian blinds to "make the window a solar collector." Notice that the black blinds will not make a single whit more energy come through the windows. In fact, from across the street you would probably be able to see the surface of those black blinds, meaning that less energy was being retained in the room. Second, the blinds will get hot, putting that high temperature right next to the window where more heat will be lost to the outside. If you think about it, no more energy gets into the house with the blinds than without them; so they really are of no help at all. A less apparent problem is this: Holding high temperatures right next to that glass can cause the same problems experienced in solar collectors that do not have special solar glazing—the glass can break.

CUTTING YOUR HEATING BILL BY 50% WITHOUT A SOLAR FURNACE

Perhaps it is heresy to discuss such a thing, but maybe you don't have the money or the inclination to buy a solar furnace right now.

For a relatively small investment, you can cut your heating bill by 50% simply by improving the weathertightness and insulation of your home.

I know. I know. Solar energy is exciting. Insulation is boring. But even if you are going to rush out and buy a solar furnace tomorrow, the very first thing you should do is improve your home to solar-utilization standards.

RULE NUMBER 23

It always costs less to upgrade your home's insulation to the necessary standards than to buy the additional solar collector equipment to make up for not doing so.

Just what are "solar standards"? R-40 insulation in your attic, R-20 insulation in your walls, storm windows and doors, adequate weatherstripping and caulking. Sounds simple, doesn't it?

Well, again, common sense is necessary. R-20 insulation in the walls is easy to obtain if you are building a new house. But the house you are living in now may only have a 3-inch batt of glass fiber in the walls. That's only an R-10. It costs a ton of money to blow insulation into the outside walls. If you have that much insulation, you may decide not to add more.

If you are building a new house, you will find that there are two points of view about the best way to obtain the desired R-20 insulation in the side walls.

One group will advocate something called the "Arkansas plan." Here, you use 2-by-6 construction instead of the traditional 2-by-4 construction, keeping the cost about equivalent by framing the house on 24-inch centers instead of on 16-inch centers. This permits the installation of a full 6 inches of glass fiber in the walls. In theory and in practice, the Arkansas plan works.

Another group maintains that the practice of framing on 24-inch centers makes the house flimsier, more susceptible to damage, because the drywall inside the house is not adequately supported on such long centers. Further, the builders in this group maintain that the framing crews are not used to working with 2-by-6 construction, and so the labor costs go up. This group also points out that the wood studs used in the side walls are not as good insulating value as the insulation going between the studs, and that, therefore, a thermal picture of the wall would show increased heat loss wherever a stud was placed.

The alternative method this group uses to achieve the same R-20 insulation is simple. They frame the house in the traditional way. The framing crews are very familiar with that. They use full-width (3-5/8 inch) aluminum-foil-faced glass-fiber batts between the studs. Instead of, or in addition to, the black insulating board used on the outside of the framing, they use a one-inch tongue-and-groove polystyrene foam insulation. Besides wrapping the entire house in a sheet of insulation, the tongue-and-groove material seals the house very well against infiltration by the outer air. Many builders also install yet another air-and-vapor barrier, a film of polyethylene or other thin plastic behind the drywall.

In your attic, you should bring the level of insulation up to at least R-40. Because insulations do not always get installed with the precision that is exercised in testing laboratories, you should install about 18 inches of glass-fiber batts, about 14 inches of rock wool or cellulose, 24 inches of vermiculite, 10 inches of polystyrene foam (bead board) or 6 inches of polyurethane foam.

If your house has the standard 4:12 roof pitch, the fibrous-glass batt will be your least expensive and, in my opinion, best insulating material. The first batt, the one nearest the ceiling, should be foil-faced. If you do a bit of looking, you should be able to find batts that don't even have the kraft-paper facing. If you can't find unfaced batts, the facing on all batts except the bottom layer should be slashed with a knife a half-dozen times to permit the migration of moisture through the batts. When upgrading attic insulation, be very sure that the attic has adequate ventilation.

If you have cathedral ceilings, flat roofing or some other architectural problem, you will have to study the available insulating materials and your house, and then use your common sense to achieve an R-40 standard.

Obviously, if your home is already built, it is impossible to add vapor barriers behind the drywall or panelling without destroying it; however, a couple of simple tricks can achieve nearly the same result.

Take a caulking gun with clear caulk and seal the space between the walls and the floor on all the outer walls. If you have wall-to-wall carpeting, simply release a small section (about 3 feet) from the tack-strip holding it to the floor. Lift the carpeting back and caulk, then reattach the carpet to the tack-strip. If you have a baseboard, caulk both above and below it.

Next, take a roll of 3-inch-wide masking or drafting tape and a razor blade or mat knife. Turn off the circuit breakers for the house. Then, one at a time, remove all the switch plates and electrical outlet covers throughout the house. Apply a piece of the tape over the whole outlet, after removing the cover. Make sure it adheres firmly to the wall and to the outlet. Replace the cover, screwing it into place. Then, using the razor-knife, trim around the outside of the plate and inside the holes for the plugs or switches, removing any of the tape that is still visible. When you have finished, you should not be able to see that the outlet has been taped. If you don't believe this hassle is worth the bother, the next time the wind blows outside put your hand in front of an outlet on an outside wall in your home. You will be able to feel the breeze.

The final step in sealing your house better requires some common sense: Caulking requires putting the caulk where it will stop air from moving into or out of your house.

KINDS OF CAULKING

There are three basic kinds of caulking material suitable for use on your home: butyl, silicone or acrylic. They all will have about the same life. If you are not the handyman type, acrylic caulking is probably the one to get. If you are using it outside, you must be fairly certain it won't rain for a week or so, because acrylic caulk is water-based and takes that long to cure to a point where it won't be dissolved. But it is a heck of a lot easier to work with than the other two caulks; you can use a damp rag to make a filled crack look neat, and clean-up is a lot easier, too.

Be sure to buy a big enough batch of caulking, regardless of the kind you get. Windows are poorly caulked, typically. Siding—where it laps over the foundation, where it joins a chimney or where it joins a doorframe—generally has a crack you could drive a truck through. You'll see, when you start looking.

OUTSIDE AIR FOR COMBUSTION

This sounds complicated, but isn't as bad as it sounds. Using ducting, provide outside air for combustion for your fueled furnace, gas-fired hot-water heater and gas-fired clothes dryer. This will reduce the amount of air filtering into your house, and also save you money, by not removing heated air from the house to provide oxygen for combustion but instead using outside air for that purpose. If you have a fireplace, you should provide outside combustion air for it, too.

VENT FANS

If you have a kitchen range hood that vents cooking odors to the outside of the house, change it. Vent the air through a charcoal filter and back into the house. Not only does that vent take heat you have paid for from the range and shoot it outside, but also it pulls heated air from near the ceiling and discharges that, too.

The same is true for bathroom vent fans. Run the air through a filter and back into the room. (Provide a small trap-tray for condensed humidity.)

Vent fans are interesting because they show how little we are conditioned to think of conservation. Only a prodigal society could design machines to blow heat out of a home in the winter time!

THE CLOTHES-DRYER

If you have an *electric* clothes dryer (but *not* if you have a gas-fired one), vent the hot-air exhaust to the inside of the house in the wintertime. Block the hole to the outside carefully, so that cold air won't blow into the laundry room. A simple, if inelegant, method to prevent the lint from blowing into the basement is to attach an old nylon stocking over the end of the air-exhaust hose. Works great, and it's easy to change.

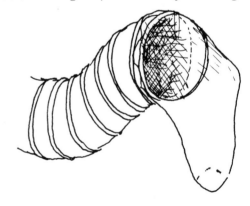

TINT FILMS

You may be reading ads for something called "tint films" that say they "reduce winter heating bills, reduce summer cooling bills." These really are great for summertime use wherever you need air conditioning. If you fall in that category, get the kind that can be used in the summer but removed in the winter.

What about their assertion that they reduce heating bills? That's a really marginal claim, and it is my opinion that it is deceptive. If you add the film to the window, it will increase the insulating value of the window a little bit. But, as we saw, a tremendous amount of energy gets into the house in the daytime through the windows. These films are designed specifically to *prevent* up to 70% of that energy's getting into the house. If you leave your drapes open all night, or if you have no drapes at all on your windows, the statement that the tint films reduce your heating bill is true. If you close your drapes at night, the statement is questionable.

If you have delicate furnishings which would be damaged by sunlight you might wish to install tint films to protect them; but you should be aware that you are going to pay a penalty in heating costs.

DEFLECTORS

The next thing you can do to reduce your heating bill is to make up for the downright looniness of some architects and builders. If you have forced-air heating, where are the registers? In nine out of ten houses they are under the windows and close to the wall, so that if the drapes are closed they blow air *behind* the drapes! If you have baseboard radiators, guess what? Under the windows, behind the drapes.

You can remedy these situations by fashioning simple deflectors out of card-board covered with aluminum foil. With registers, tape them into place so that the air is directed toward the center of the room instead of blowing up behind the curtains. For radiators, install the foil reflector on the wall directly above the radiator and extending about 5 inches into the room. Shorten drapes wherever necessary, too, to keep them from covering the radiators.

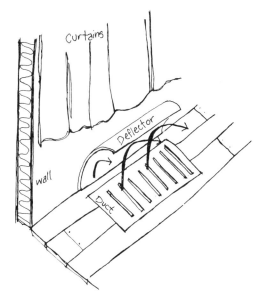

CONTINUOUS AIR CIRCULATION

If you have forced-air heating in your home now, you can save from 15% to 30% on your heating bill, probably without buying a thing. All that is necessary is to convert your furnace to Continuous Air Circulation (C.A.C.). All that means is that the furnace fan is slowed down somewhat and set to run continuously, 24 hours a day, during the heating season.

Right now, with forced-air heat from registers, when your house thermostat calls for heat the first thing that happens is that the gas flame is turned on. The furnace then heats up until the heat exchanger gets to about 85° F. During that time the fan is not on. A lot of heat is lost up the chimney during this stage. Then when it gets to 85° the fan turns on and runs until it reaches about 150° F. At that point the flame is turned off. The fan continues to run until the heat exchanger is cooled to 85° again. As we have already seen, the hotter something is, the faster it loses heat through insulation. Well, if you put 150° air next to the ceiling it is going to lose heat more rapidly than if the air were only 75°. This system makes the furnace a very rapid response unit. But you have air blowing into the house from the registers at all temperatures between 85° and 150°. And are surprised when the heat is uneven and feels drafty. (Simply because of this surging of heat, many people will specify hot-water heat using baseboard radiators.)

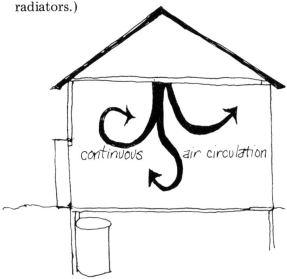

continuous air circulation

With continuous circulation, that airflow pattern from the registers is changed. The system is quieter because the fan is running more slowly. Less heat is lost up the chimney, since the fan is already running when your thermostat turns the gas on. You have less draftiness, and the temperature of the delivered air typically is much

cooler—more in the range of 80° to 90° F. Generally, it doesn't cost more electricity to run the fan all the time at the slower speed than it costs to run it with starts and stops: Starting takes a bit more energy than running and besides, because less air is being moved with the continuous running, it takes less energy per minute.

There is one drawback: The heat delivered is much more even, like hot water radiant heating; but, just like hot-water heating, the response time of the system is slow. For example, if you have the kitchen door open for a quarter of an hour while unloading groceries (unlikely to take that long at today's prices!), the house may take that long again to regain the thermostat setting. Just as with hot-water radiators.

The reason a lot of heating contractors don't tell you about continuous air circulation, even though the furnace they are installing is equipped for it, is this: If you have been used to the on/off air system, you are used to the quick response time. More important, if you hold your hand over the register of an on/off unit the air delivery at higher temperatures feels very warm to the hand. But with C.A.C., that air will generally only be about 80° F. Your hand is nearly 98.6° F. So the air will feel cool to your hand. Many times the homeowner, not realizing that this 80° air can heat the home quite adequately, will think that the furnace is malfunctioning and will call the furnace dealer out on an unnecessary service call. Many dealers are of opinion that it is easier to install the furnace in the old-fashioned way than to try to educate the customer.

You can convert the furnace to C.A.C. yourself, if you are fairly handy. If not, figure on paying your local heating contractor about $15 to $20 to come out and do the conversion.

STORM WINDOWS AND DOORS FOR $20

If you don't have storm windows and doors you are going to have to be very careful about purchasing them. You need storm windows and doors even if you live in the Deep South. Not only will they cut your heating bills; they will reduce your cooling bills, as well. There is a host of rip-off artists selling these at about five times more than you need to pay. Try to find a local, reputable contractor who does not employ direct-sales people on straight commission.

What if you are on a limited or fixed income and can't afford the cost of storm windows and doors? There's a way for you to do the whole job yourself for $20 or even less.

What you do is simple, if tedious. You go to the hardware store and buy transparent thin-film plastic. Make sure that it is clear, not milky. Buy some giant-sized rolls of ordinary transparent tape in the one-inch width. Get some plastic reinforced with metal screening for all the screen doors, and several boxes of thumbtacks. Get weather stripping for all doors *and* screen doors. Here's what you do:

Cut the thin clear plastic sheeting to the exact size that will permit it to cover the panes and lap the window sash about 1/2 inch all the way around. Tape it to the sash, not to the frame, so that the window can be opened. Do this on the *inside.* Do *not* cover the window from the outside of the house, the way farmers used to. The plastic will discolor, get brittle, and probably won't survive the winter. Inside, with care, it will last several years.

You generally need two sets of hands to get the film attached to the sash without tell-tale wrinkles showing. If you can't get a helper, tape the top first, then the bottom, then the sides. Be sure the film is sealed all the way around the window. If the window sash permits a choice of places to tape the plastic, choose the one that will leave between the plastic and the glass an air space as large as possible, but not more than one inch.

If you do this job carefully, the storm window you have just made will function almost as well as the regular storm sash. I would advise starting with basement or bedroom windows, leaving kitchen and living room windows to the last, since there is probably going to be a real improvement in your workmanship, with the experience.

Now for the doors. Cut the reinforced plastic to size and tack into place, covering the entire screened area of the door. Keep the plastic as taut as possible. Then add weather stripping to the screen door as well as to the main door. You may have to increase the screen door's spring tension to make it self-closing with the weather stripping in place. If you do a careful job, you will have a very good storm door.

If you have a mail slot that goes through the door or through the wall, stuff it full of insulation and caulk it shut. Replace it with an outside mailbox.

BATHS AND SHOWERS

If you take showers, put the stopper in the tub before you begin to keep the water from running out of the tub. If you take a bath, leave the water standing in the tub when you finish. The water will be at 100° F or so when you finish your shower or bath. Let the water stand in the tub until it is cold (room temperature). Leave the bathroom door open while the tub is cooling. It is amazing, but that full tub of water, in cooling from 100° to 70° F, will give up enough energy to keep a small properly insulated three-bedroom house warm for one full hour when it is 10° above zero—Fahrenheit—outside! You've already paid good money for that heat energy; you might as well use it instead of letting it run down the drain. One tip, though. If you are going to do this, put some bubble bath or a little bit of detergent in the water. (Anything that contains a wetting agent.) Otherwise, you get a ring that will be hard to remove.

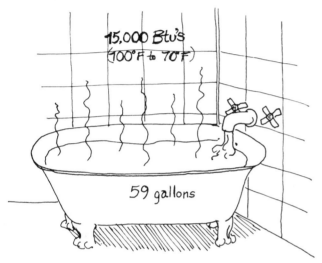

Similarly, if on a lesser scale, the hot water you use for washing dishes should be allowed to cool to room temperature before you let it drain. If your washing machine drains into a laundry tub, do the same thing.

As you might guess, it is a real energy saver to wash your clothes in cold water with one of the new cold-water soaps. If you haven't tried any of these, do so. You may be surprised at the job they do. And your clothes will last longer, shrink less and fade less.

YOUR PORCH LIGHT

Another simple one. If you use your outside porch light regularly at night, take a look at both your porch and your interior floor plan: There may very well be a way to use

a bright light *inside* the house that will shine through the window to illuminate porch and front sidewalk. A 100-watt light bulb will emit more than 4,000 Btu's of heat over the nighttime hours. If the lamp is outside, that energy is lost; if it is inside, it will reduce your heating bill. When you turn out the light at night, don't forget to close the drapes on that window.

FIDDLING WITH THE THERMOSTAT

If you have a frame house, even one with brick veneer, you should fiddle with the thermostat. If your house is solid masonry, you probably should not. Turn the thermostat down to 65° F at night as you go to bed. If that doesn't make getting up the next morning miserable, turn it down to 60° the next night. Find the lowest temperature that doesn't make the next morning an ordeal. Turn the thermostat back up next morning. Turn it down when you go to work. This will save a batch of energy: On the order of 6.75 million Btu's over the winter in a well-insulated small three-bedroom house. More in bigger houses. Several manufacturers make a timed thermostat that does this automatically for you. It costs on the order of $50 to $60 and will pay for itself in one or two heating seasons.

CATHEDRAL CEILINGS AND CLERESTORY WINDOWS

Both cathedral ceilings and clerestories are mammoth energy thieves. You pay a real penalty for having this extra space you can't live in. It is extremely difficult to make up for this poor design. Adding insulation to what invariably seems to be an inadequate amount on cathedral ceilings is always difficult and always expensive.

Try adding some circulation fans to push the hot air from near the ceiling toward the floor to help cut the heat loss. Figure

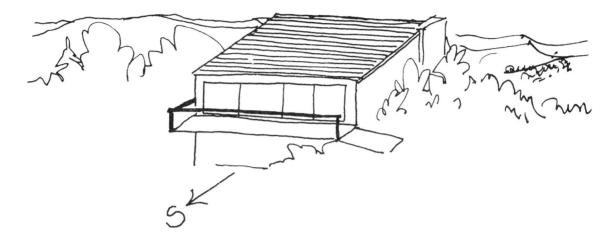

out some way to drape the clerestories with heavily lined drapes to cut heat loss from them at night. Better yet, board them up and insulate them.

Obviously, if you are building a new house, don't let the architect design any clerestory windows or cathedral ceilings into the plans. I know, you'll probably have to get into a wrestling match to prevent them. If you want the "wide-open feeling" of a cathedral ceiling, try a two-foot elevation of the living-room ceiling or a sunken conversation area.

A little knowledge is a dangerous thing, and many architects have become enraptured by a concept called the "passive solar home." Generally, what they do is build a ton of windows on the south side of the house. Don't let your architect do that to your home. Probably a better name for this concept is "uncontrolled solar overheating." If you get too much glass on the south side of the house, you are going to make it miserable.

For example, a 40-foot window wall on the south side of a house in Minneapolis would transmit 200,000 Btu's to the inside of the house on an average January day, taking cloudiness into account. More than

300,000 on a sunny day. Yet that house only needs about 150,000 Btu's during the daytime to keep it at 70° F. (A 7,000 DD, or Degree-Day, house.) You will have two problems. First, the house will get uncomfortably warm and, second, the rooms will be uncomfortably bright. You can't read a book, for example, without closing some of the drapes. You would probably be better off, in the long run, to build the house with the ordinary amount of windows distributed on all sides of the house.

Clerestory windows

THE FIREPLACE

Your fireplace can really be a villain when it comes to stealing heat from your home. The best of chimney dampers provides a poor seal, so that when a fire isn't going there is a continual loss of heated air up the chimney and out of the house. If you don't believe it, walk over to the fireplace and put your hand under the damper. A miniature windstorm.

The solution is simple. Put glass doors on your fireplace. Despite your initial negative reaction to that, the glass doors don't really have to take away the fun of a fireplace. The doors' frames have a small draft control at the bottom. By judicious use of this control, your logs will last a heck of a lot longer than they do when burning uncontrolled. The heat still comes into the room through the glass. You can still hear the fire. More important, you are no longer pulling a lot of heated air from the house to provide combustion air in unlimited amounts, nor are you simply losing heated air to the heavy draft that blows up the chimney when the fire is burning. As stated earlier, you should provide ducts to bring combustion air from outside to the fireplace. But if you don't, the extra draft control will help.

MAKING YOUR OWN INSULATION

You can make an insulation similar to one which is sold under the euphemistic name, "cellulose." That is nothing more than shredded newspaper that has been soaked

in a fire-retardant chemical. If you are short of cash but need that additional insulation in your attic, all you need to do is start saving the newspapers.

Discard the glossy sheets and comics. Keep just the regular pulp-paper parts of the newspaper. Using a standard cheese grater that you pick up at the dime store, shred the paper into small pieces. Store the shreddings in a box. When you have collected several boxfuls, you are ready for the mess of fire-retardant treatment.

In a washtub, following the manufacturer's instructions, mix a batch of fire-retardant chemical. Two of the most readily available are *Minalith* and *Pyresote*. If they aren't available from the local chemical house, a good substitute will be. The chemical formulas are, "diammonium phosphate, ammonium sulfate, sodium tetraborate and boric acid," and, "zinc chloride, ammonium sulfate, boric acid and sodium bichromate," respectively, for the two. The major ingredient is plain old boric acid. These fire retardants do *not* make the shredded newspaper, or "cellulose," fire *proof*. They simply make it less combustible. The same as the "cellulose insulation" you can buy. Be sure, however, to follow the manufacturer's instructions for mixing the chemicals. Generally, you will be mixing one pound of chemicals to one gallon of water and soaking the pulp for 15 minutes. Fortunately, the fire-retardant chemicals are not expensive. You will be able to treat enough "cellulose" for insulating your whole attic to a depth of 12 inches for about 50 bucks plus your own elbow grease.

To dry the wet shredded paper, the easiest thing is to spread it thinly over more newspapers laid out on a table or in the back yard, or on a piece of screening salvaged from an old screen door. *Do not install the insulation while it is damp.* It will have a gray appearance, not unlike rock wool. When it is totally dried, you can either store it in a garbage bag or install it directly in the attic. Depending on how easy it is to get up there, and how many creaks in the bones.

Use your common sense. Do not install the "cellulose" too close to light fixtures that are recessed and extend up into the attic. Stay about a foot away from any of those. If you have an attic furnace, stay well away from the furnace and take the time to tack some garbage bags or leaf bags over the "cellulose" nearest to the furnace so that winter winds blowing through the attic vents won't blow the material against the furnace. Remember that you have only treated it with a fire-*retarding* material, not a fire-*proofing* material.

When using the cheese grater, roll the newspaper tightly, and go to work. Use the coarsest grate on the grater. This is *not* easy. Don't expect to use the grater for cheese or carrots again. But, if you haven't got the bucks to buy insulation, this is a good way to go. Tip: Do the grating while watching TV, since it is really tedious. But watch the fingers!

SECTION THREE

Selecting Your Dealer, Buying Solar Stocks and Determining the Degree-Day Size of Your House

8 Selecting the Right Dealer

PRICE OR EXPERIENCE?

Now comes the really hard part. One of the biggest problems you are going to face when adding sun-power-using equipment to your home is the fact that the industry is so young.

Let's say you have carefully checked out the various solar furnaces for sale in your community and you find one that fits all the rules in this book. The equipment passes muster. Even so, the installation can be rotten. Not on purpose. Most of the dealers in this new industry are very sincere. But sincerity can't make up for a lack of experience or training.

So, unlike other purchases, where the price oftentimes is the single determining factor in a buying decision, the buying of solar equipment requires a long, hard look at the dealer who will perform the installation.

If you are of an impatient temperament, don't even consider buying solar equipment from a brand-new dealer. Find a dealer who has installed several systems already and is willing to refer you to several customers who are satisfied with their job.

Why? Because there is a corollary to Murphy's Law that states, "Even if it can't be hooked up wrong, the new dealer will find a way." The new dealer, even with the best of training, will make some mistakes. They won't be of the irretrievable sort. After all, a solar furnace is little more than a light-trap and an insulated box or tank. It's pretty hard to mess those up. Rather, the mistakes will be of the irritating sort. Things like forgetting to tie your fueled furnace into the controller properly. So that the solar unit works fine through September and October, but late in November is unable to supply all the heat the house needs. Then the solar unit shuts off, but your fueled furnace doesn't turn on. And it happens at 8:00 on Friday evening. And the dealer has gone hunting for the weekend. That kind of frustration. Only takes a minute to fix, if you know what you're doing.

Then, another thing happens with a dealer's first installation. It is something I call the "great mystery syndrome." Because this is "solar" equipment, and this is "new technology," no mere mortal can possibly understand the equipment. It is pretty hard to conjure up a mystery in a box of rocks; but for some reason, when something goes wrong in that first installation the dealer will lose all hold on reason and turn the malfunction into the deepest mystery possible. For example, a usual first-time mistake is to hook up the sensors for the collector temperature and storage temperature backward on the controller. This means that the collector will not turn on until it is 10° *colder* than the storage temperature, instead of the reverse, which is proper. So the collector doesn't turn on until nighttime, when the collector cools to the outside temperature. As the inventor of one system, I got a call from a dealer one morning at 2:00 A.M. He thought the *moon* was turning on the collector. A great mystery!

All this is by way of preparing you for the worst if you buy the first unit a dealer ever installs. He or she may learn very quickly. She or he most certainly has attended an intensive factory training school in order to become a dealer. But somehow that classroom won't substitute for experience.

Therefore, if you are the guinea pig, the first customer, and you are willing to put up with the learning, you have every right to insist upon a $400 or $500 discount (the dealer profit). For irritation.

COMMONSENSE PRECAUTIONS
The other things you should do fall into the area of standard procedure.

For example, check out the dealer, *before you buy*, with the local Better Business Bureau, or, if you are fortunate enough to live in a city where they exist, the District Attorney's Consumer Protection Agency.

Check the warranty that comes with the solar unit extremely carefully.

Unless you are a professional engineer with a few years' experience in engineering solar heating systems, don't believe that you can buy collectors and engineer the system yourself. In my opinion, it is extremely deceptive to give you the idea you can simply buy an El Supremo solar collector, build a rock bin in your basement or buy a water tank, buy some fans or pumps, and hook it all together to get an efficient, cost-effective system. Believe me, some of the most disastrous systems in the country were assembled from components in that way by engineers. The manufacturer's literature makes it sound so easy.

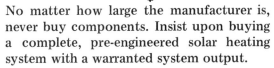

RULE NUMBER 24
No matter how large the manufacturer is, never buy components. Insist upon buying a complete, pre-engineered solar heating system with a warranted system output.

It just stands to reason. If the manufacturer is confident about the product, there is absolutely no reason why he or she can't engineer a complete system for you and stand behind the *system* performance. Remember what we learned earlier: A collector is the easiest and least expensive part to manufacture of the whole solar-heat-using system. For example, one manufacturer who sells a complete system for a suggested retail price of $27 per square foot of collector sells the collector alone to commercial accounts for large buildings at $6 per square foot. In fact, you can figure that the collector accounts for only about a quarter to a fifth of the cost of the entire system. I think it is outrageous that some of the manufacturers sell you collector *alone* for $15 or $20 per square foot with minimal insulation (1 inch to 3 inches of fibrous glass), and then expect you to do the engineering, buy all the rest of the components, and try to hook it all together to work efficiently.

Installation costs can vary significantly from house to house, particularly when the equipment is being added to an existing house. Be sure to get a written estimate of installation costs before signing on the dotted line.

If the system is pre-engineered and designed with you in mind, it will conform to the *existing* building codes in your home town. Make sure that the building permit (which *is* necessary) was obtained without petitions for variances from local building codes. (Sometimes, because of covenants on a property, variances will be necessary before you add a structure. It is not this kind of variance I am talking about.)

The reason this is important is simple. The building code was written for your protection. It is a set of standards written to protect your health and safety. For example, the collector covers should be of tempered glass. Some manufacturers will use ordinary glass and then piously proclaim to local authorities that special "solar building codes" are necessary. Nonsense. They simply want to save a few bucks at the risk of your safety. Surely you don't want a child falling into a collector and getting cut because the manufacturer didn't build with safety glass.

There is another possibility of problems if you put a nonconforming structure on your house: Some insurance policies state that nonconforming structures will void the protection. It would be a real shock to have your house burn down and then find out the insurance company wasn't going to pay off because some manufacturer saved a couple of bucks instead of simply building the solar unit to code.

So, before you put the solar furnace on your house, make sure it conforms to code. Then tell your insurance agent that you are planning to install solar equipment. Find out if it is going to raise your homeowner's policy rates.

It has been my experience that, if you install a system that is designed to conform to building codes and is installed by a trained professional, the addition of solar equipment will not raise your premium, in most cases.

Many states offer incentives for solar-energy utilization and, by the time this book is in print, it is a good bet that there will be federal income-tax incentives for installing solar equipment on your home. Make sure before you buy that the equipment you are purchasing qualifies for these incentives.

9 Solar Stocks

YOU, TOO, CAN GET RICH QUICK

I'll probably make a lot of enemies with this very short chapter.

If I were you, for a few years yet I wouldn't buy any stock in companies purporting to offer equipment for transforming the sun's energy for domestic power and heating uses. Unless you are going to buy enough to gain control of the company, or you are personal friend to the management.

There are enough sleazy practices going on right now in the solar-energy area to curl the hair on any S.E.C. investigator's head.

I have seen 10-cent stocks sold for 30, 40 and even 50 times their value, by any kind of evaluation. Even an optimistic future projection. Some of these "penny stocks" are selling for such inflated values that even if the issuing company were to sell 500-million dollars' worth of solar equipment in a year, they could not pay a 10% dividend. At a time when one company's stock was selling at one price, they sold nearly a half-million dollars' worth of additional stock at a price almost 20 times less than the trading price to some private investors. The other stockholders evidently didn't object! I have seen a company's stock drop in price nearly 10 times on the two-year anniversary date of the offering. That is the day when the original stockholders are permitted to sell their founder's stock.

I have seen one company run through nearly half a million dollars in less than half a year and close their doors. Yet that stock went up by nearly 300% the week after President Carter's energy message!

Colorado is the place where we had a uranium boom a number of years ago. At that time, the offerings were for stock at a penny a share. Inflation has raised it to a dime.

RULE NUMBER 25

Unless you got rich trading in penny uranium stocks, don't dabble in solar-energy stocks, at least until 1980.

Actually, some of the 10-cent stocks will probably turn out to be winners in the future. I have trouble envisioning how, but anything is possible. Most of the companies are undercapitalized, have inexperienced business and production management, and pay their executives as much as long-established successful companies.

If you are bound and determined to invest in solar-energy stocks, here is a pretty good rule of thumb: See how many shares are issued and outstanding in the company, and see how much you are going to have to pay per share; multiply those two numbers together; divide the answer by 0.35. That will give you a fairly good number for how much their sales have to be to pay you a 10% dividend each year. You ought to get 10% on your investment. You are running more risk (!!!) and you can get 6% at the bank, where it is safe.

There are sound companies, without doubt, in the solar-energy business, and they may warrant investment. It probably is still too soon, however, except for the inveterate gambler.

10 Performing a Heat-Loss Calculation on Your Home

USING THE DEGREE-DAY HOME-SIZER
This is the single most important thing you can do when you are considering lowering your home-heating bill or when you are planning to add solar heating to your house.

RULE NUMBER 26
Do not even shop for solar heating equipment until you have determined the degree-day size of your home.

You must know the Degree-Day size of your house in order to know how much solar equipment you need.

It is extremely simple to find out the Degree-Day size of your house. All you need is a tape measure and about half an hour. Although you can do the multiplying by hand, a pocket calculator is extremely handy for performing the simple multiplications that will be necessary.

WHAT IS A "DEGREE DAY"?
The term "Degree Day" may be new to you; so let's take a minute to understand it.

The base of the Degree-Day system is 65° Fahrenheit. The assumption is that absolutely no heat is required by a structure being kept at 70° Fahrenheit if the temperature outside is 65° or warmer. If the *average* temperature over a 24-hour period is 64° F, that is *one* Degree Day; if the average temperature is 63° F for 24 hours, that is *two* Degree Days, et cetera.

For many years, the Weather Bureau has kept very careful records of the average temperatures, month by month, in a number of towns. The Bureau then publishes tables of normal Degree Days for those towns. By using these tables, it is possible to estimate very closely the fuel consumption of a given building. Although the numbers obtained by such estimates may not be correct for a given day, they can be quite accurate when taken over the entire 273-day heating season.

THE DEGREE-DAY HOUSE
Although a solar furnace delivers Btu's to the house, the fuel it uses—sunshine—is intermittent as a supply source. Some days

are cloudy enough so that no collection takes place. Therefore, it becomes fairly impractical to use a solar furnace as a 100% supply source. It is much more economical to depend upon it for anywhere from 40% to 90% of the home's heating requirements and to use it only as an auxiliary heating source. A fueled or electric furnace is used for the main system.

This means that a new kind of engineering must be done—a kind called "average-case" engineering.

Now, we can calculate fairly accurately the amount of heat needed by a house at any given outside temperature. This calculation is stated, usually, in terms of the Btu's required per hour to keep the house at 70° F, when it is that given temperature outside.

When we are going to do average-case engineering on a particular house and already know, from Weather Bureau measurements, the average sunshine available in various towns and cities, we also need to know the *average* heat required by that house during the heating season. A new term has been invented for this purpose: the "Degree-Day Home Size." Using this approach, the home is measured in terms of the Btu's needed for each Degree Day.

For example, a 5,000-Degree-Day house would be one that would need 5,000 Btu's during a 24-hour period when the outside temperature was 64° F. If the outside

temperature were 30° F for 24 hours, that same house would need 175,000 Btu's during that 24-hour period in order to keep the house at 70° F.

MEASURING YOUR HOUSE

Using a tape measure, determine the dimensions of all the windows and doors in your house. Measure all the outside walls. HINT: It is really helpful to draw a rough sketch of your home's floor plan and write the dimensions on that sketch.

Multiply the height by the width to obtain the square footage of your windows, doors, outside walls and floor, etc.

Now, using the worksheet provided and the "Degree-Day Home-Sizer" below, perform the calculations indicated.

When you have finished with the multiplication of the factors, add all the products together. That is the Degree-Day size of your home. It's that simple.

HOW DOES YOUR HOUSE STACK UP?

Finally, a comparison table has been provided on page 75 (Table 2) of various types of houses and the approximate Degree-Day size they should be if properly insulated. Find the nearest description to yours and see how well it stacks up. If the Degree-Day size of your home is far afield from the values listed in the comparison table, you should turn your attention to upgrading your insulation.

Table 1   DEGREE-DAY HOME SIZER

Instructions: First, find the closest description to the construction of your house. Then take the appropriate multiplier, times the square footage of your home's outside walls, windows, ceilings, floors, etc., to obtain the Degree-Day size of your home.

OUTSIDE WALLS

FRAME CONSTRUCTION WITH SIDING OR BRICK VENEER

	Multiplier		*Multiplier*
No insulation whatsoever	8.90	Full-width batt with foil facing	1.75
No insulation, foil facing on		Full-width batt plus 1-inch styrene foam	1.45
inside of walls	6.05	Full-width batt with foil plus 1-inch	
1-inch batt	3.85	styrene foam	1.35
2-inch batt	2.65	6-inch batt	1.20
3-inch batt	2.05	6-inch batt with foil facing	1.15
Full-width batt (3-5/8")	1.90	6-inch batt with foil plus 1-inch	
Rock wool filling entire cavity	1.90	styrene foam	0.95

LOG CONSTRUCTION—SQUARED, TONGUE AND GROOVE, WELL SEALED

	Maple or Oak—Wall Thickness				Pine or Fir—Wall Thickness			
	6-Inch	8-Inch	10-Inch	12-Inch	6-Inch	8-Inch	10-Inch	12-Inch
Multiplier	4.60	3.45	2.75	2.30	3.20	2.40	1.90	1.60

MASONRY CONSTRUCTION—MULTIPLIERS

Type	Thickness (in inches)	Wall Alone, No Finish on Inside Wall	1/2" Plaster or Drywall Inside	2" Furring, 1/2" Plaster or Drywall Inside	2" Furring 1" Batt, 1/2" Plaster, Drywall
Brick	8 inches	12.00	11.05	7.45	3.35
	12 inches	8.65	8.15	6.00	3.10
	16 inches	6.70	6.50	5.05	2.90
Stone	8 inches	16.80	15.35	9.10	3.85
	12 inches	13.60	12.70	8.15	3.60
	16 inches	11.75	10.80	7.45	3.35
	24 inches	8.90	8.40	6.25	3.10
Poured Concrete	6 inches	18.95	17.05	9.85	3.85
	8 inches	16.80	15.35	9.10	3.85
	10 inches	15.10	13.90	8.65	3.60
	12 inches	13.70	12.70	8.15	3.60
Concrete Block	8 inches	13.45	12.50	8.15	3.60
	12 inches	11.75	11.05	7.45	3.35
Cinder Block	8 inches	9.85	9.35	6.70	3.10
	12 inches	9.10	8.65	6.25	3.10

HOLLOW GLASS BLOCKS
(7-3/4″ x 7-3/4″ x 3-7/8″): Smooth—11.75
Ribbed—11.05

ESTIMATING MULTIPLIER FOR CONCRETE WALLS UNDERGROUND: 2.60

WINDOWS AND OUTSIDE WALLS

Single Pane—27.10   Double Pane—10.80   Triple Pane—6.75

CEILINGS

Heated Floor Above	0
No insulation whatsoever	16.00
2-inch batt, fibrous glass, mineral wool	2.95
2 inches vermiculite	4.35
2 inches rock wool or cellulose	2.80
2 inches polystyrene (bead board)	2.50
4-inch batt	1.60
4 inches vermiculite	2.55
4 inches rock wool or cellulose	1.55
4 inches polystyrene (bead board)	1.35
6-inch batt	1.10
6 inches vermiculite	1.80
6 inches rock wool or cellulose	1.05
6 inches polystyrene	0.95
10-inch batt	0.70
10 inches vermiculite	1.10
10 inches rock wool or cellulose	0.65
10 inches polystyrene	0.60
12-inch batt	0.60
12 inches vermiculite	0.95
12 inches rock wool or cellulose	0.55
12 inches polystyrene	0.50
14-inch batt	0.50
18-inch batt	0.40

OUTSIDE DOORS

Width of Door (in inches)	Open to Outside	With Glass Storm Door
1 inch	16.55	10.10
1-1/4 inches	14.15	9.10
1-1/2 inches	12.50	8.40
1-3/4 inches	12.25	8.40
2 inches	11.05	7.70
2-1/2 inches	9.10	6.70

FLOORS

Concrete slab	2.40
Over uninsulated crawl space	3.70
Over crawl space,	
4-inch batt insulation	1.40
6-inch batt	0.95
10-inch batt	0.60
12-inch batt	0.50
Over unheated basement—use appropriate "crawl space" multiplier	
Over heated basement	0
Don't forget to compute basement walls and basement floor, however.	
Basement floor	2.40

Notes to Degree-Day Home-Sizer: *(1) Degree-Day Home Size = Heat Loss of Home in Btu's per Degree Day. (2) Assumes no heat lost from home at outside ambient temperatures above 65° F. (3) Value of k (Btu/ft^2· hr·°F) used to obtain multiplier may be found by dividing multiplier by 24 (hours).*

WORKSHEET FOR DETERMINING DEGREE-DAY HOME SIZE

ALL WINDOWS	(A) TOTAL SQUARE FOOTAGE: LENGTH TIMES WIDTH		(B) MULTIPLIER FROM TABLE 1		(C) TOTAL OF COLUMN A TIMES COLUMN B
ALL DOORS		X		=	
OUTSIDE WALLS		X		=	
CEILING		X		=	
FLOOR		X		=	
BASEMENT WALLS		X		=	
MISCELLANEOUS		X		=	
TOTAL COLUMN C = HOME SIZE IN DEGREE DAYS					

USE THE GRID BELOW FOR A ROUGH SKETCH OF YOUR HOUSE AND TO INDICATE DIMENSIONS

WORKSHEET FOR DETERMINING DEGREE-DAY HOME SIZE

ALL WINDOWS	(A) *TOTAL SQUARE FOOTAGE: LENGTH TIMES WIDTH*		(B) *MULTIPLIER FROM TABLE 1*		(C) *TOTAL OF COLUMN A TIMES COLUMN B*
ALL DOORS		X		=	
OUTSIDE WALLS		X		=	
CEILING		X		=	
FLOOR		X		=	
BASEMENT WALLS		X		=	
MISCELLANEOUS		X		=	
TOTAL COLUMN C = HOME SIZE IN DEGREE DAYS					

USE THE GRID BELOW FOR A ROUGH SKETCH OF YOUR HOUSE AND TO INDICATE DIMENSIONS

WORKSHEET FOR DETERMINING DEGREE-DAY HOME SIZE

	(A) *TOTAL SQUARE FOOTAGE: LENGTH TIMES WIDTH*		(B) *MULTIPLIER FROM TABLE 1*		(C) *TOTAL OF COLUMN A TIMES COLUMN B*
ALL WINDOWS					
ALL DOORS		X		=	
OUTSIDE WALLS		X		=	
CEILING		X		=	
FLOOR		X		=	
BASEMENT WALLS		X		=	
MISCELLANEOUS		X		=	

TOTAL COLUMN C = HOME SIZE IN DEGREE DAYS

USE THE GRID BELOW FOR A ROUGH SKETCH OF YOUR HOUSE AND TO INDICATE DIMENSIONS

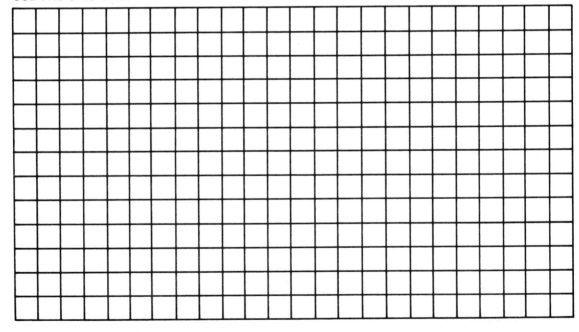

WORKSHEET FOR DETERMINING DEGREE-DAY HOME SIZE

	(A) TOTAL SQUARE FOOTAGE: LENGTH TIMES WIDTH		(B) MULTIPLIER FROM TABLE 1		(C) TOTAL OF COLUMN A TIMES COLUMN B
ALL WINDOWS					
ALL DOORS		X		=	
OUTSIDE WALLS		X		=	
CEILING		X		=	
FLOOR		X		=	
BASEMENT WALLS		X		=	
MISCELLANEOUS		X		=	
TOTAL COLUMN C = HOME SIZE IN DEGREE DAYS					

USE THE GRID BELOW FOR A ROUGH SKETCH OF YOUR HOUSE AND TO INDICATE DIMENSIONS

WORKSHEET FOR DETERMINING DEGREE-DAY HOME SIZE

ALL WINDOWS	(A) *TOTAL SQUARE FOOTAGE: LENGTH TIMES WIDTH*		(B) *MULTIPLIER FROM TABLE 1*		(C) *TOTAL OF COLUMN A TIMES COLUMN B*
ALL DOORS		X		=	
OUTSIDE WALLS		X		=	.
CEILING		X		=	
FLOOR		X		=	
BASEMENT WALLS		X		=	
MISCELLANEOUS		X		=	

TOTAL COLUMN C = HOME SIZE IN DEGREE DAYS

USE THE GRID BELOW FOR A ROUGH SKETCH OF YOUR HOUSE AND TO INDICATE DIMENSIONS

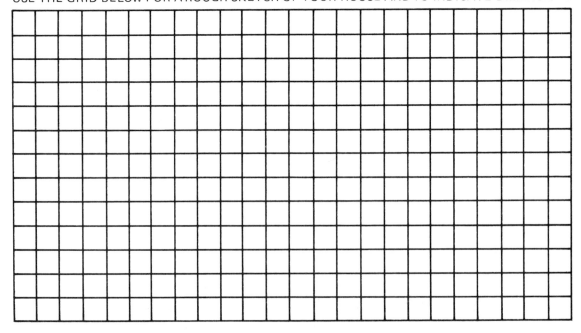

Table 2 COMPARISON TABLE—APPROXIMATE DEGREE-DAY SIZES OF PROPERLY INSULATED HOUSES

Instructions: After you have completed the calculations of the Degree-Day size of your house, compare the result with the table below. If your house has a higher Degree-Day number than the one found below, you need to add insulation. Note: In determining the square footage of your house, count above-ground area only. Do not include basement square footage.

Square Footage of House	NEW CONSTRUCTION			EXISTING HOUSES		
	Slab Floor	Crawl Space or Unheated Basement	Heated Full Basement	Slab Floor	Crawl Space or Unheated Basement	Heated Full Basement
RANCH-STYLE HOUSES						
750	4700	3300	7200	5200	3800	7700
1000	5800	3900	8700	6400	4500	9300
1500	7900	5000	11400	8600	5700	12100
2000	9800	6000	13900	10600	6800	14700
2500	11800	7000	16300	12700	7900	17200
3000	13600	7900	18600	14600	8900	19600
3500	15500	8900	20900	16600	10000	22000
4000	17400	9800	22800	18500	10900	24000
TWO-STORY HOUSES						
1000	5000	4100	7000	5900	4900	7900
1500	6500	5100	9000	7500	6100	10000
2000	7800	5900	10700	9000	7100	11900
2500	9200	6800	12400	10500	8100	13800
3000	10400	7500	13900	11900	9000	15400
3500	11700	8400	15500	13300	10000	17100
4000	12900	9100	17000	14600	10800	18700

TRI-LEVEL HOUSES

Square Footage of House	New Construction	Existing Houses	Square Footage of House	New Construction	Existing Houses
1000	5300	6000	3000	10600	11800
1500	6700	7600	3500	11900	13100
2000	8000	9000	4000	13000	14400
2500	9400	10500			

Notes to Comparison Table: *(1) Assumptions: 750 ft² house—125 ft² windows, 42 ft² doors; 1,000 ft² house—150 ft² windows, 42 ft² doors; 1,500 ft² house—175 ft² windows, 63 ft² doors, 2,000 ft² house—200 ft² windows, 63 ft² doors; 2,500 ft² house—225 ft² windows, 84 ft² doors; 3,000 ft² house—250 ft² windows, 84 ft² doors; 3,500 ft² house—275 ft² windows, 105 ft² doors; 4,000 ft² house—300 ft² windows, 105 ft² doors. New houses: R-40 attic, R-17 side wall, R-40 crawl space, R-6.67 basement walls, concrete basement floors. Existing houses, same, except R-12 side walls. (2) Values used obtained from Table 1.*

SECTION FOUR

Determining the Size of Solar Collector Needed for Heating Both Your Home and the Water Your Family Uses

11 Estimating the Size of Collector Needed for Your House

USING THE SOLAR-COLLECTOR-GRAM
Now you know the Degree-Day size of your home. As part of the determination of that number, you compared the Degree-Day size of your house with a table of what it should be. If your house needs more insulation, go back and redo the calculations, using the multipliers that will be appropriate after you have added the insulation.

It cannot be emphasized strongly enough that the proper insulation of your home is the single most important, and certainly the most cost-effective, home improvement you can make.

To use the Solar-Collector-Gram (Nomogram 1), which will give you a good estimate of the extent of collector surface necessary for your home, take the Degree-Day size of your house times the multiplier for your home town from the table of multipliers, Table 4. If your home town isn't listed, take the nearest city that approximates the winter conditions in your town.

NOTE: If you live in the mountains, and your home town is not listed, find the nearest plains city and use the appropriate additive from Table 3, below, for *each* 1,000 feet additional altitude of your home town above the plains city.

Table 3 ALTITUDE CORRECTION ADDITIVES FOR USE WITH SOLAR-COLLECTOR-GRAM MULTIPLIERS

Average Btu's Per Day	Additive Per 1,000 Ft. Alt.
300- 307	10.0
308- 324	9.5
325- 342	9.0
343- 363	8.5
364- 387	8.0
388- 413	7.5
414- 444	7.0
445- 480	6.5
481- 521	6.0
522- 571	5.5
572- 631	5.0
632- 705	4.5
706- 800	4.0
801- 923	3.5
924-1090	3.0
1091-1333	2.5
1334-1500	2.0

Note to Altitude Correction Table: *Assume 3°F drop in temperature for each 1,000 foot increase in altitude.*

In order to know which additive to use, you will have to refer to the tables of "Daily Average Btu Output From Solar Furnace" (Table 11) on page 92. Find the Btu number for the plains city. Then refer to Table 3 to find the appropriate additive. Once you have found the right additive, add it to the multiplier for the plains city. Example: The multiplier for the plains city is 20. Your home town is 1,500 feet higher than the plains city and 20 miles away. The average Btu output for the plains city is 792. You would add 6 to the multiplier (4 x 1-1/2 = 6), which would be 26 (20 + 6 = 26).

Now go directly to the Solar-Collector-Gram. Using a straightedge, line up the number obtained by taking the multiplier times the Degree-Day size of your house (on the right) with the percentage of heat over the winter season that you desire (on the left). Where the straightedge crosses the center line is the approximate amount of solar collector your home would need.

Remember that this is only an estimate. It will be within 10% one way or the other, however.

If you want to satisfy yourself that the manufacturers who say things like, "It takes a collector from 15% to 30% of the square footage of your home to heat it," are full of beans, take the time to try several different Degree-Day house sizes on the Solar-Collector-Gram.

If you want more concise sizing estimates, go to Appendix D, where worksheets and instructions are provided so that you can compute how large a collector is needed for your home.

It is important to note that this Solar-Collector-Gram is based on the assumption that you are going to be purchasing a unit with two covers, with R-30 insulation on the sides and back and a horizontal reflector at the base of the collector, and with a storage system with R-30 insulation.

If, for some unknown reason, you are planning to purchase a collector and storage unit with less insulation and lesser specifications, do not use this Solar-Collector-Gram. Instead, you should calculate the Btu output using the worksheets provided in Chapter Thirteen and those provided in Appendix D.

Nomogram 1 SOLAR-COLLECTOR-GRAM
(For determining approximate size of solar collector needed for your house)

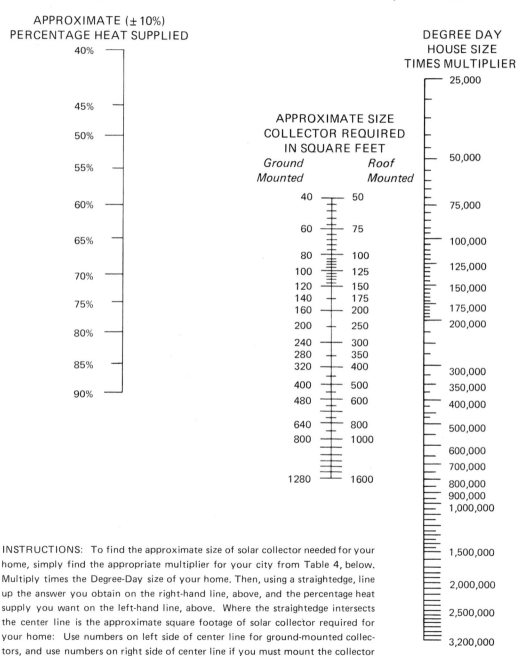

APPROXIMATE (±10%)
PERCENTAGE HEAT SUPPLIED

DEGREE DAY
HOUSE SIZE
TIMES MULTIPLIER

APPROXIMATE SIZE
COLLECTOR REQUIRED
IN SQUARE FEET

Ground Mounted   *Roof Mounted*

INSTRUCTIONS: To find the approximate size of solar collector needed for your home, simply find the appropriate multiplier for your city from Table 4, below. Multiply times the Degree-Day size of your home. Then, using a straightedge, line up the answer you obtain on the right-hand line, above, and the percentage heat supply you want on the left-hand line, above. Where the straightedge intersects the center line is the approximate square footage of solar collector required for your home: Use numbers on left side of center line for ground-mounted collectors, and use numbers on right side of center line if you must mount the collector on the roof.

Table 4 SOLAR-COLLECTOR-GRAM MULTIPLIERS FOR PROPERLY DESIGNED
SOLAR UNITS (R-30-INSULATED COLLECTOR, DOUBLE COVER, WITH
HORIZONTAL REFLECTORS, R-30-INSULATED STORAGE)(SEE TABLES 5
THROUGH 10 FOR OTHER SYSTEMS.)

	Btu Output for 60°-Tilt Collector	Btu Output for 90°-Tilt Collector		Btu Output for 60°-Tilt Collector	Btu Output for 90°-Tilt Collector
ALABAMA			Grand Junction	29.5	30.0
Birmingham	17.5	18.0	Pueblo	21.0	22.0
Huntsville	21.0	22.0	CONNECTICUT		
Mobile	9.5	10.0	Bridgeport	34.0	35.5
Montgomery	14.0	14.5	Hartford	38.5	39.0
ARIZONA			New Haven	31.5	32.5
Flagstaff	26.0	28.0	DELAWARE		
Phoenix	7.0	7.0	Wilmington	29.0	30.0
Prescott	16.5	17.0	DISTRICT OF COLUMBIA		
Tucson	7.0	7.0	Washington	25.0	26.0
Winslow	20.0	21.0	FLORIDA		
Yuma	5.0	4.5	Appalachicola	7.0	7.5
ARKANSAS			Daytona Beach	4.5	6.5
Fort Smith	22.5	23.0	Jacksonville	7.5	8.0
Little Rock	22.0	22.5	Key West	0.8	0.7
Texarkana	16.0	16.0	Miami Beach	1.0	1.0
CALIFORNIA			Orlando	4.0	3.5
Bakersfield	10.0	10.0	Pensacola	9.5	9.5
Bishop	18.5	19.0	Tallahassee	9.0	9.0
Burbank	6.0	6.5	Tampa	4.0	4.5
Eureka	21.5	23.0	GEORGIA		
Fresno	12.5	13.0	Athens	17.5	17.5
Long Beach	6.0	6.0	Atlanta	18.0	18.5
Los Angeles	6.5	7.0	Augusta	14.5	15.0
Oakland	12.0	12.5	Columbus	14.0	14.0
Red Bluff	14.5	15.0	Macon	11.0	11.5
Sacramento	15.0	15.5	Rome	20.5	21.0
San Diego	5.0	5.0	Savannah	9.5	10.0
San Francisco	12.0	12.5	IDAHO		
Santa Maria	9.5	10.5	Boise	35.0	35.0
COLORADO			Idaho Falls	60.0	62.0
Alamosa	42.0	44.0	Lewiston	46.0	46.5
Colorado Springs	26.0	26.5	Pocatello	50.0	50.5
Denver	27.0	27.5	ILLINOIS		

	Btu Output for 60°-Tilt Collector	Btu Output for 90°-Tilt Collector		Btu Output for 60°-Tilt Collector	Btu Output for 90°-Tilt Collector
Cairo	22.0	23.0	MASSACHUSETTS		
Chicago	40.0	41.5	Boston	34.0	34.0
Moline	45.0	47.0	Nantucket	32.0	34.0
Peoria	38.0	39.0	Pittsfield	55.0	57.5
Rockford	47.0	48.0	Worcester	42.0	44.0
Springfield	34.5	35.0	MICHIGAN		
INDIANA			Alpena	63.5	64.0
Evansville	30.5	33.0	Detroit	50.5	51.5
Fort Wayne	44.0	45.0	Escanaba	56.5	58.5
Indianapolis	38.0	39.0	Flint	56.0	58.0
South Bend	48.0	50.0	Grand Rapids	56.0	57.0
IOWA			Lansing	55.0	55.5
Burlington	35.0	36.0	Marquette	66.0	68.5
Des Moines	40.0	41.0	Muskegon	53.5	54.0
Dubuque	47.5	50.0	Sault Ste. Marie	75.0	77.5
Sioux City	40.0	41.0	MINNESOTA		
Waterloo	45.5	47.5	Duluth	62.5	64.0
KANSAS			International Falls	78.5	80.0
Concordia	28.0	29.0	Minneapolis	53.5	55.0
Dodge City	22.0	26.0	Rochester	51.5	52.0
Goodland	27.0	28.0	Saint Cloud	60.0	61.5
Topeka	31.0	31.5	MISSISSIPPI		
Wichita	22.5	25.5	Jackson	15.5	16.0
KENTUCKY			Meridian	15.0	15.0
Covington	35.0	36.0	Vicksburg	13.0	13.5
Lexington	34.0	37.5	MISSOURI		
Louisville	32.5	35.0	Columbia	31.0	31.5
LOUISIANA			Kansas City	27.0	27.0
Alexandria	11.5	12.0	St. Joseph	33.0	33.0
Baton Rouge	9.5	9.5	St. Louis	31.0	31.0
Lake Charles	8.5	8.5	Springfield	27.0	27.0
New Orleans	8.0	8.0	MONTANA		
Shreveport	12.5	13.0	Billings	39.0	41.5
MAINE			Glasgow	56.5	57.5
Caribou	86.5	91.5	Great Falls	38.5	40.0
Portland	36.5	39.5	Havre	55.5	58.0
MARYLAND			Helena	47.5	50.0
Baltimore	27.0	27.0	Kalispell	60.5	61.5
Frederick	32.0	33.5	Miles City	45.0	47.5

	Btu Output for 60°-Tilt Collector	Btu Output for 90°-Tilt Collector		Btu Output for 60°-Tilt Collector	Btu Output for 90°-Tilt Collector
Missoula	60.5	61.5	Winston-Salem	22.5	23.0
NEBRASKA			NORTH DAKOTA		
Grand Island	35.0	36.0	Bismarck	34.0	35.0
Lincoln	32.5	33.0	Devils Lake	36.5	38.0
Norfolk	38.5	40.0	Fargo	35.5	38.0
North Platte	32.5	34.0	Williston	35.0	35.0
Omaha	36.5	37.5	OHIO		
Scottsbluff	31.5	32.5	Akron	52.5	53.0
NEVADA			Cincinnati	30.0	30.0
Elko	34.0	35.5	Cleveland	53.5	54.5
Ely	32.0	34.0	Columbus	40.0	40.5
Las Vegas	11.0	11.0	Dayton	34.5	34.5
Reno	23.0	24.0	Mansfield	48.5	49.5
Winnemucca	29.0	31.0	Sandusky	44.0	45.0
NEW HAMPSHIRE			Toledo	30.0	31.0
Concord	47.5	50.0	Youngstown	32.5	33.0
NEW JERSEY			OKLAHOMA		
Atlantic City	25.0	26.5	Oklahoma City	18.5	19.0
Newark	29.0	29.5	Tulsa	21.0	21.5
Trenton	29.5	30.0	OREGON		
NEW MEXICO			Astoria	36.0	37.5
Albuquerque	15.5	16.5	Eugene	36.0	37.0
Raton	24.5	25.5	Medford	36.0	37.0
Roswell	15.0	14.5	Pendleton	41.5	42.5
Silver City	14.5	15.0	Portland	36.5	37.5
NEW YORK			Roseburg	36.0	36.5
Albany	47.5	49.5	Salem	38.5	39.5
Binghampton	56.5	58.0	PENNSYLVANIA		
Buffalo	56.5	58.0	Allentown	32.5	33.5
New York City	28.5	29.5	Erie	54.5	56.5
Rochester	55.0	57.5	Harrisburg	30.5	31.0
Schenectady	47.5	48.0	Philadelphia	27.5	27.5
Syracuse	58.5	60.0	Pittsburgh	48.0	48.5
NORTH CAROLINA			Reading	31.5	32.0
Asheville	21.0	21.5	Scranton	43.5	44.0
Charlotte	16.0	16.0	Williamsport	46.5	47.0
Greensboro	19.5	20.0	RHODE ISLAND		
Raleigh	17.5	17.5	Providence	31.5	32.0
Wilmington	13.0	13.5	SOUTH CAROLINA		

	Btu Output for 60°-Tilt Collector	Btu Output for 90°-Tilt Collector		Btu Output for 60°-Tilt Collector	Btu Output for 90°-Tilt Collector
Charleston	10.5	10.5	VIRGINIA		
Columbia	15.0	15.5	Lynchburg	21.5	22.0
Florence	14.5	14.5	Norfolk	19.0	19.5
Greenville	17.5	18.0	Richmond	20.5	21.0
Spartanburg	19.0	19.5	Roanoke	29.0	29.5
SOUTH DAKOTA			WASHINGTON		
Huron	48.5	49.5	Olympia	38.0	38.5
Rapid City	34.5	35.5	Seattle	34.0	35.0
Sioux Falls	46.5	47.5	Spokane	52.0	55.0
TENNESSEE			Tacoma	39.0	40.0
Bristol	26.5	26.5	Walla Walla	39.5	40.0
Chattanooga	20.5	21.0	Yakima	38.0	38.0
Knoxville	21.5	22.0	WEST VIRGINIA		
Memphis	19.5	20.0	Charleston	38.0	38.5
Nashville	22.5	23.0	Elkins	39.5	42.0
TEXAS			Huntington	35.0	35.0
Abilene	12.5	12.5	Parkersburg	40.0	41.5
Amarillo	15.0	15.0	WISCONSIN		
Austin	12.5	12.5	Green Bay	57.0	61.0
Brownsville	4.5	4.5	La Crosse	55.0	57.0
Corpus Christi	6.0	6.5	Madison	56.0	58.0
Dallas	13.5	13.5	Milwaukee	49.0	52.0
El Paso	11.0	11.0	WYOMING		
Fort Worth	13.5	13.5	Casper	34.5	36.0
Galveston	7.5	7.5	Cheyenne	32.0	33.5
Houston	8.5	9.5	Lander	35.0	36.0
Laredo	5.0	5.0	Sheridan	39.0	40.0
Lubbock	16.0	16.0	CANADA		
Midland	11.0	11.5	ALBERTA		
Port Arthur	9.0	9.0	Banff	102.0	108.0
San Angelo	10.0	10.5	Beaverlodge	100.0	101.0
San Antonio	9.5	9.5	Brooks	77.0	78.0
Victoria	7.0	7.0	Calgary	70.0	72.0
Waco	12.0	12.5	Edmonton	78.0	80.0
Wichita Falls	16.5	16.5	Edson	90.0	92.5
UTAH			Fairview	124.5	124.0
Salt Lake City	33.0	33.5	Lacombe	80.0	85.0
VERMONT			Lethbridge	62.5	64.0
Burlington	66.5	68.5	Medicine Hat	75.0	76.0

	Btu Output for 60°-Tilt Collector	Btu Output for 90°-Tilt Collector		Btu Output for 60°-Tilt Collector	Btu Output for 90°-Tilt Collector
Olds	75.0	77.5	NOVA SCOTIA		
Vauxhall	70.0	71.0	Annapolis Royal	68.0	70.0
BRITISH COLUMBIA			Halifax	50.0	50.5
Campbell River	61.0	64.0	Kentville	70.5	71.5
Dawson Creek	110.0	115.0	Sydney	65.5	67.5
Fort St. John	116.5	118.5	Truro	67.5	68.5
Hope	84.0	86.0	Yarmouth	59.5	59.5
Kamloops	96.5	98.5	ONTARIO		
Kitimat	103.5	104.0	Atikokan	95.0	96.0
Nanaimo	59.5	60.0	Belleville	71.0	74.0
Nelson	91.5	92.5	Brampton	65.0	67.5
New Westminster	61.0	61.5	Chatham	66.0	68.0
Penticton	79.0	80.5	Cornwall	74.0	75.0
Port Alberni	63.5	64.0	Guelph	74.0	75.0
Powell River	62.0	62.0	Hamilton	60.5	62.0
Prince George	108.5	110.0	Harrow	67.5	69.0
Prince Rupert	91.5	93.5	Kapuskasing	124.5	126.5
Quesnel	101.0	102.5	Kingston	62.5	63.5
Summerland	86.5	88.5	Lindsay	82.5	83.5
Terrace	108.5	110.0	London	77.0	77.5
Trail	113.5	115.0	Mount Forest	84.0	85.0
Vancouver	59.5	60.5	New Liskeard	112.5	116.0
Victoria	47.5	47.5	North Bay	84.0	85.0
MANITOBA			Ottawa	75.5	77.0
Brandon	95.0	98.0	St. Catharines	74.5	77.0
Dauphin	90.0	92.5	Sarnia	65.0	66.5
Morden	86.0	90.0	Sault Ste. Marie	84.0	86.5
Rivers	86.0	90.0	Thunder Bay	78.5	80.5
The Pas	129.0	131.0	Toronto	61.5	62.0
Winnipeg	91.5	93.5	Woodstock	75.0	76.5
NEW BRUNSWICK			PRINCE EDWARD ISLAND		
Campbellton	76.5	77.5	Charlottetown	73.5	75.0
Chatham	67.0	68.0	QUEBEC		
Fredericton	58.5	59.5	Amos	113.5	116.0
Moncton	64.0	65.0	Berthierville	76.0	78.5
Saint John	46.5	47.5	Grandes Bergeronnes	86.0	86.5
NEWFOUNDLAND			L'Assomption	78.5	81.5
Gander	95.0	98.5	Lennoxville	91.0	91.5
St. John's	84.0	88.0	Maniwaki	89.5	90.5

	Btu Output for 60°-Tilt Collector	Btu Output for 90°-Tilt Collector		Btu Output for 60°-Tilt Collector	Btu Output for 90°-Tilt Collector
Montreal	70.0	70.5	Indian Head	92.5	95.5
Normandin	107.5	109.0	Melfort	116.5	118.5
Quebec	92.5	93.5	Moose Jaw	91.0	92.5
Rimouski	106.5	108.0	Outlook	99.0	100.5
Ste. Agathe-des-Monts	86.5	87.5	Prince Albert	117.5	118.5
St. Ambroise	106.5	108.5	Regina	101.5	103.5
Sept-Iles	79.0	80.5	Saskatoon	98.5	100.0
Sherbrooke	83.5	85.0	Scott	116.0	116.0
Victoriaville	83.5	85.0	Swift Current	86.5	87.5
SASKATCHEWAN			Weyburn	84.5	86.0
Estevan	73.5	75.0	Yorkton	106.5	108.5

Notes to Table 4, Solar-Collector-Gram Multipliers: *(1) R-30 Insulation on sides and back of collector, R-30 insulated above-ground storage, double-cover collector with reflectors at horizontal. (2) Methodology described in Appendix D utilized to obtain amount of collector required to supply varying percentages of heat over 273-day heating season to various Degree-Day home sizes in the cities listed. (3) No variances greater than ± 10% found between calculated and nomogram values. (4) Ground-mounted collectors' increase in transmitted energy due to albedo inputs assumed to be 20%, although in real world increases of 5% to 35% will be seen, depending upon size of yard and upon ground reflectance conditions.*

COMPARISON TABLES FOR POORLY DESIGNED SOLAR EQUIPMENT

INSTRUCTIONS: If you wish to compare poorly designed equipment with R-30-insulated equipment with double covers on the collector and horizontal reflectors, the tables below will permit you to see how performance compares.

First, use the Solar-Collector-Gram to determine the square footage of well-designed solar collector necessary for your home. Then find, in the tables below, the equipment you wish to compare. The number you find there, when multiplied by the square footage of well-designed equipment, will give you the footage of collector the poorly designed system would need to deliver equal heating.

Example: You have determined that 200 square feet of collector are required to provide 75% of your home's heating requirements in a low-temperature system. However, a major national company has been advertising a pretty rooftop collector with one plastic cover and 2-1/2 inches of fiber glass insulation on the collector, with 2-1/2 inches of fiber glass insulation on a storage tank inside the house. No reflector is provided. You would multiply 200 x 1.7 (from Table 7 below) = 340 square feet of collector.

If the same "pretty" system were to be used to heat hot water for baseboard radiators, you would use the high-temperature table (Table 10). The "N/A" means that the system would not work under typical winter conditions—"Not Applicable."

It is here that the importance of proper insulation becomes apparent.

USE THESE TABLES FOR LOW-TEMPERATURE SYSTEMS
(TYPICAL OPERATION BETWEEN 75° AND 120° F)

Table 5   SYSTEMS WITH OUTSIDE, ABOVE-GROUND STORAGE

	AMOUNT OF INSULATION ON SIDES AND BACK OF COLLECTOR AND SURROUNDING STORAGE					
	R-3.33, 1" Fibrous Glass	R-6.67, 2" Fibrous Glass	R-8.33, 2-1/2" Fibrous Glass	R-10, 3" Fibrous Glass	R-15, 2" Urethane Foam	R-30, 4" Urethane Foam
Single Cover Without Reflectors	Not/App	7.0	3.7	2.8	2.0	1.5
Single Cover With Reflectors	Not/App	5.0	2.7	2.0	1.4	1.1
Double Cover Without Reflectors	Not/App	5.9	3.1	2.3	1.6	1.2
Double Cover With Reflectors	Not/App	4.6	2.5	1.9	1.3	1.0

Table 6   SYSTEMS WITH OUTSIDE, UNDERGROUND STORAGE

	AMOUNT OF INSULATION ON SIDES AND BACK OF COLLECTOR AND SURROUNDING STORAGE					
	R-3.33, 1" Fibrous Glass	R-6.67, 2" Fibrous Glass	R-8.33, 2-1/2" Fibrous Glass	R-10, 3" Fibrous Glass	R-15, 2" Urethane Foam	R-30, 4" Urethane Foam
Single Cover Without Reflectors	25.7	2.5	2.1	1.9	1.6	1.3
Single Cover With Reflectors	17.8	1.8	1.5	1.3	1.2	1.0
Double Cover Without Reflectors	21.0	2.1	1.8	1.6	1.3	1.1
Double Cover With Reflectors	16.0	1.6	1.4	1.2	1.1	0.9

Table 7   SYSTEMS WITH INSIDE-THE-HOUSE STORAGE

	AMOUNT OF INSULATION ON SIDES AND BACK OF COLLECTOR AND SURROUNDING STORAGE					
	R-3.33, 1" Fibrous Glass	R-6.67, 2" Fibrous Glass	R-8.33, 2-1/2" Fibrous Glass	R-10, 3" Fibrous Glass	R-15, 2" Urethane Foam	R-30, 4" Urethane Foam
Single Cover Without Reflectors	3.4	1.8	1.7	1.6	1.4	1.3
Single Cover With Reflectors	2.3	1.3	1.2	1.1	1.0	1.0
Double Cover Without Reflectors	2.7	1.5	1.4	1.3	1.2	1.1
Double Cover With Reflectors	2.1	1.2	1.1	1.0	0.9	0.9

USE THESE TABLES FOR HIGH-TEMPERATURE SYSTEMS
(TYPICAL OPERATION BETWEEN 160° AND 200° F)

Table 8 SYSTEMS WITH OUTSIDE, ABOVE-GROUND STORAGE

	AMOUNT OF INSULATION ON SIDES AND BACK OF COLLECTOR AND SURROUNDING STORAGE					
	R-3.33, 1" Fibrous Glass	R-6.67, 2" Fibrous Glass	R-8.33, 2-1/2" Fibrous Glass	R-10, 3" Fibrous Glass	R-15, 2" Urethane Foam	R-30, 4" Urethane Foam
Single Cover Without Reflectors	N/A	N/A	N/A	N/A	N/A	N/A
Single Cover With Reflectors	N/A	N/A	N/A	N/A	50.6	10.1
Double Cover Without Reflectors	N/A	N/A	N/A	N/A	8.6	2.8
Double Cover With Reflectors	N/A	N/A	N/A	N/A	5.9	2.0

Table 9 SYSTEMS WITH OUTSIDE, UNDERGROUND STORAGE

	AMOUNT OF INSULATION ON SIDES AND BACK OF COLLECTOR AND SURROUNDING STORAGE					
	R-3.33, 1" Fibrous Glass	R-6.67, 2" Fibrous Glass	R-8.33, 2-1/2" Fibrous Glass	R-10, 3" Fibrous Glass	R-15, 2" Urethane Foam	R-30, 4" Urethane Foam
Single Cover Without Reflector	N/A	N/A	N/A	N/A	N/A	N/A
Single Cover With Reflector	N/A	N/A	N/A	146.4	30.1	9.0
Double Cover Without Reflector	N/A	N/A	N/A	35.7	5.1	2.5
Double Cover With Reflector	N/A	N/A	N/A	23.6	3.5	1.8

Table 10 SYSTEMS WITH INSIDE-THE-HOUSE STORAGE

	AMOUNT OF INSULATION ON SIDES AND BACK OF COLLECTOR AND SURROUNDING STORAGE					
	R-3.33, 1" Fibrous Glass	R-6.67, 2" Fibrous Glass	R-8.33, 2-1/2" Fibrous Glass	R-10, 3" Fibrous Glass	R-15, 2" Urethane Foam	R-30, 4" Urethane Foam
Single Cover Without Reflector	N/A	N/A	N/A	N/A	N/A	N/A
Single Cover With Reflector	N/A	N/A	185.0	164.4	24.2	8.5
Double Cover Without Reflector	N/A	N/A	77.6	11.3	4.1	2.4
Double Cover With Reflector	N/A	N/A	49.2	7.4	2.8	1.7

Notes: (1) Based on Tables 16-27. (2) For purposes of average performance comparison, outside temperature assumed to be 30° F.

Nomogram 2 SOLAR-COLLECTOR-GRAM FOR DOMESTIC WATER HEATERS
(For determining how large the collector should be for various family sizes)

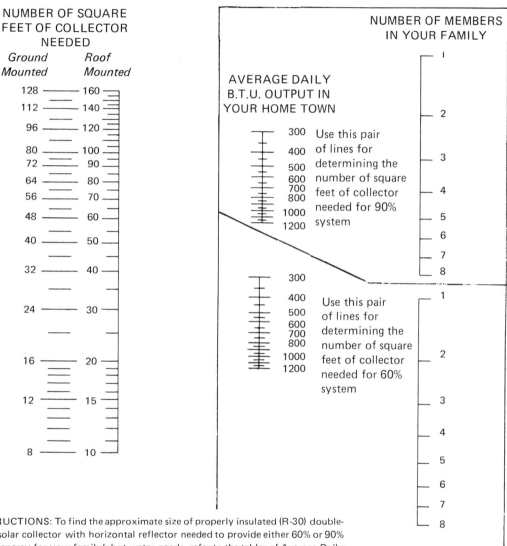

INSTRUCTIONS: To find the approximate size of properly insulated (R-30) double-cover solar collector with horizontal reflector needed to provide either 60% or 90% of the energy for your family's hot-water needs, refer to the tables of Average Daily Btu Output in Table 11, below, for the appropriate number for your city. Locate that number on the center line. (Use the upper portion for a 90% supply, the lower for a 60% supply.) Using a straightedge, line that point up with the point on the right-hand side corresponding to the number of members in your family. Where the straightedge intersects the left-hand line you will find the approximate number of square feet of collector needed to provide that percentage of your domestic hot water.

Notes to Solar-Collector-Gram for Water Heaters: (1) Assumes 25 gallons per day used by first family member, 15 gallons per day by each additional family member. (2) Water temperature, 140°F; inlet temperature year-round assumed to be 55 F. (3) Mean Absorber Temperature ($T_{in} + T_{out}/2$) = 140 F. (4) R-30 insulation on water storage tank. Air/water heat exchange and air collector assumed, with pebble-bed storage.

Table 11 DAILY AVERAGE B.T.U. OUTPUT FROM ONE SQUARE FOOT OF R-30-INSULATED DOUBLE-COVER SOLAR COLLECTOR TAKEN OVER 273-DAY HEATING SEASON (SEPTEMBER 1 THROUGH MAY 31)

	Btu Output for 60°-Tilt Collector	Btu Output for 90°-Tilt Collector		Btu Output for 60°-Tilt Collector	Btu Output for 90°-Tilt Collector
ALABAMA			Grand Junction	842	808
Birmingham	750	718	Pueblo	920	887
Huntsville	737	710	CONNECTICUT		
Mobile	783	741	Bridgeport	684	657
Montgomery	812	770	Hartford	654	630
ARIZONA			New Haven	733	705
Flagstaff	885	852	DELAWARE		
Phoenix	1133	1088	Wilmington	724	696
Prescott	1024	987	DISTRICT OF COLUMBIA		
Tucson	1176	1127	Washington	720	692
Winslow	1002	965	FLORIDA		
Yuma	1224	1172	Appalachicola	858	810
ARKANSAS			Daytona Beach	845	697
Fort Smith	748	721	Jacksonville	820	776
Little Rock	776	748	Key West	923	841
Texarkana	822	791	Miami Beach	886	818
CALIFORNIA			Orlando	869	810
Bakersfield	1002	965	Pensacola	853	809
Bishop	888	852	Tallahassee	845	798
Burbank	996	960	Tampa	893	832
Eureka	619	594	GEORGIA		
Fresno	966	928	Athens	804	771
Long Beach	994	959	Atlanta	770	738
Los Angeles	990	955	Augusta	826	788
Oakland	878	843	Columbus	814	778
Red Bluff	913	877	Macon	815	812
Sacramento	907	871	Rome	729	698
San Diego	944	909	Savannah	811	771
San Francisco	850	818	IDAHO		
Santa Maria	943	913	Boise	724	699
COLORADO			Idaho Falls	606	585
Alamosa	723	695	Lewiston	603	582
Colorado Springs	852	822	Pocatello	573	552
Denver	842	812	ILLINOIS		

	Btu Output for 60°-Tilt Collector	Btu Output for 90°-Tilt Collector		Btu Output for 60°-Tilt Collector	Btu Output for 90°-Tilt Collector
Cairo	784	753	MASSACHUSETTS		
Chicago	667	641	Boston	670	644
Moline	649	625	Nantucket	694	667
Peoria	692	665	Pittsfield	563	543
Rockford	652	628	Worcester	624	601
Springfield	724	695	MICHIGAN		
INDIANA			Alpena	506	488
Evansville	725	645	Detroit	573	550
Fort Wayne	629	602	Escanaba	557	539
Indianapolis	667	639	Flint	550	529
South Bend	612	588	Grand Rapids	548	526
IOWA			Lansing	562	540
Burlington	756	727	Marquette	466	449
Des Moines	711	684	Muskegon	546	525
Dubuque	636	611	Sault Ste. Marie	459	442
Sioux City	733	700	MINNESOTA		
Waterloo	689	664	Duluth	568	549
KANSAS			International Falls	548	529
Concordia	818	787	Minneapolis	619	598
Dodge City	873	787	Rochester	639	617
Goodland	857	826	Saint Cloud	604	583
Topeka	736	709	MISSISSIPPI		
Wichita	839	753	Jackson	731	694
KENTUCKY			Meridian	772	733
Covington	671	643	Vicksburg	782	742
Lexington	643	572	MISSOURI		
Louisville	688	613	Columbia	739	710
LOUISIANA			Kansas City	780	750
Alexandria	803	761	St. Joseph	746	718
Baton Rouge	786	745	St. Louis	724	695
Lake Charles	790	746	Springfield	747	719
New Orleans	767	724	MONTANA		
Shreveport	842	806	Billings	684	662
MAINE			Glasgow	653	631
Caribou	436	421	Great Falls	705	683
Portland	681	659	Havre	648	626
MARYLAND			Helena	643	623
Baltimore	731	698	Kalispell	520	502
Frederick	677	645	Miles City	703	679

	Btu Output for 60°-Tilt Collector	Btu Output for 90°-Tilt Collector		Btu Output for 60°-Tilt Collector	Btu Output for 90°-Tilt Collector
Missoula	527	509	Winston-Salem	732	702
NEBRASKA			**NORTH DAKOTA**		
Grand Island	775	745	Bismark	641	620
Lincoln	767	737	Devils Lake	623	602
Norfolk	739	712	Fargo	616	596
North Platte	804	774	Williston	661	638
Omaha	779	748	**OHIO**		
Scottsbluff	796	767	Akron	557	532
NEVADA			Cincinnati	668	641
Elko	787	756	Cleveland	551	527
Ely	825	795	Columbus	636	610
Las Vegas	1079	1040	Dayton	689	661
Reno	880	845	Mansfield	589	564
Winnemucca	825	792	Sandusky	618	593
NEW HAMPSHIRE			Toledo	583	560
Concord	608	587	Youngstown	544	520
NEW JERSEY			**OKLAHOMA**		
Atlantic City	745	716	Oklahoma City	824	794
Newark	735	706	Tulsa	783	756
Trenton	727	699	**OREGON**		
NEW MEXICO			Astoria	465	449
Albuquerque	1031	936	Eugene	506	488
Raton	858	828	Medford	572	549
Roswell	939	905	Pendleton	636	614
Silver City	1027	982	Portland	503	486
NEW YORK			Roseburg	517	496
Albany	614	592	Salem	501	484
Binghampton	484	464	**PENNSYLVANIA**		
Buffalo	550	530	Allentown	685	658
New York City	735	706	Erie	532	510
Rochester	562	540	Harrisburg	681	654
Schenectady	614	592	Philadelphia	721	693
Syracuse	525	504	Pittsburgh	571	547
NORTH CAROLINA			Reading	661	635
Asheville	739	712	Scranton	590	566
Charlotte	827	797	Williamsport	556	533
Greensboro	779	750	**RHODE ISLAND**		
Raleigh	778	748	Providence	688	663
Wilmington	856	821	**SOUTH CAROLINA**		

	Btu Output for 60°-Tilt Collector	Btu Output for 90°-Tilt Collector		Btu Output for 60°-Tilt Collector	Btu Output for 90°-Tilt Collector
Charleston	869	831	**VIRGINIA**		
Columbia	818	785	Lynchburg	759	730
Florence	805	772	Norfolk	784	754
Greenville	813	783	Richmond	766	736
Spartanburg	754	726	Roanoke	661	635
SOUTH DAKOTA			**WASHINGTON**		
Huron	705	682	Olympia	490	472
Rapid City	744	720	Seattle	491	474
Sioux Falls	705	681	Spokane	575	555
TENNESSEE			Tacoma	480	463
Bristol	651	625	Walla Walla	609	588
Chattanooga	709	682	Yakima	683	660
Knoxville	705	678	**WEST VIRGINIA**		
Memphis	761	732	Charleston	571	548
Nashville	718	689	Elkins	559	536
TEXAS			Huntington	596	572
Abilene	919	880	Parkersburg	558	534
Amarillo	964	930	**WISCONSIN**		
Austin	758	715	Green Bay	604	585
Brownsville	716	652	La Crosse	635	613
Corpus Christi	799	742	Madison	602	580
Dallas	845	808	Milwaukee	620	597
El Paso	1044	994	**WYOMING**		
Fort Worth	845	808	Casper	766	740
Galveston	792	738	Cheyenne	792	763
Houston	723	682	Lander	805	778
Laredo	841	783	Sheridan	708	686
Lubbock	967	929	**CANADA**		
Midland	988	943	**ALBERTA**		
Port Arthur	770	725	Banff	376	365
San Angelo	913	868	Beaverlodge	433	420
San Antonio	753	711	Brooks	499	483
Victoria	788	733	Calgary	493	478
Waco	805	766	Edmonton	489	474
Wichita Falls	841	808	Edson	426	413
UTAH			Fairview	404	392
Salt Lake City	789	756	Lacombe	466	452
VERMONT			Lethbridge	546	529
Burlington	492	475	Medicine Hat	495	478

	Btu Output for 60°-Tilt Collector	Btu Output for 90°-Tilt Collector		Btu Output for 60°-Tilt Collector	Btu Output for 90°-Tilt Collector
Olds	480	456	NOVA SCOTIA		
Vauxhall	523	506	Annapolis Royal	416	402
BRITISH COLUMBIA			Halifax	477	462
Campbell River	381	368	Kentville	418	404
Dawson Creek	425	412	Sydney	423	409
Fort St. John	421	408	Truro	415	400
Hope	309	299	Yarmouth	441	425
Kamloops	402	389	ONTARIO		
Kitimat	305	295	Atikokan	448	433
Nanaimo	418	403	Belleville	449	434
Nelson	386	373	Brampton	494	476
New Westminster	422	407	Chatham	468	449
Penticton	439	423	Cornwall	458	442
Port Alberni	415	400	Guelph	439	423
Powell River	406	392	Hamilton	474	455
Prince George	369	357	Harrow	466	464
Prince Rupert	219	213	Kapuskasing	338	325
Quesnel	384	371	Kingston	491	473
Summerland	433	417	Lindsay	432	416
Terrace	303	293	London	435	418
Trail	350	337	Mount Forest	438	422
Vancouver	426	410	New Liskeard	354	342
Victoria	493	475	North Bay	442	426
MANITOBA			Ottawa	458	442
Brandon	474	459	St. Catharines	416	399
Dauphin	491	475	Sarnia	467	449
Morden	496	478	Sault Ste. Marie	427	411
Rivers	511	494	Thunder Bay	480	463
The Pas	410	397	Toronto	485	468
Winnipeg	506	490	Woodstock	444	427
NEW BRUNSWICK			PRINCE EDWARD ISLAND		
Campbellton	446	431	Charlottetown	421	407
Chatham	471	456	QUEBEC		
Fredericton	454	439	Amos	378	364
Moncton	460	445	Berthierville	485	468
Saint John	489	474	Grandes Bergeronnes	427	413
NEWFOUNDLAND			L'Assomption	468	452
Gander	335	324	Lennoxville	415	401
St. John's	337	327	Maniwaki	444	427

	Btu Output for 60°-Tilt Collector	Btu Output for 90°-Tilt Collector		Btu Output for 60°-Tilt Collector	Btu Output for 90°-Tilt Collector
Montreal	493	476	Indian Head	509	493
Normandin	399	385	Melfort	424	411
Quebec	406	392	Moose Jaw	509	492
Rimouski	352	340	Outlook	482	466
Ste. Agathe-des-Monts	456	439	Prince Albert	443	429
St. Ambroise	393	378	Regina	493	476
Sept-Iles	462	445	Saskatoon	517	501
Sherbrooke	434	419	Scott	447	432
Victoriaville	451	436	Swift Current	498	481
SASKATCHEWAN			Weyburn	529	512
Estevan	569	550	Yorkton	461	446

Note to Table 11, Daily Average Btu Output From One Square Foot of Solar Collector Over 273-Day Heating Season (September 1 through May 31): *Based upon values in Table 47 (Entire Heating Season Btu Output from One Square Foot of Properly Designed and Insulated Collector) divided by 273. For conditions, see notes to Table 47.*

12 The Monthly Cost of "Free" Solar Energy

DETERMINING ELECTRICAL COSTS

Although the solar energy is free, we have already seen that the equipment needed to harvest it is expensive. And—although this is disillusioning—it costs money every month, as well.

Some solar furnaces have two motors, some only have one. It costs you money to run a motor. Therefore, before you purchase a solar furnace, you should estimate the cost of *operating* that furnace.

In Table 13, below, you will find the number of hours (approximately) that the collector motor can be expected to run every month. Find that number for your home town. *Whether or not there is a separate motor on the pump or fan that delivers heat from storage to the house, the power consumption is the same;* so don't get thrown off the track by that. Ignore it.

If you have a system that supplies 50% or less of the heat required for your house during the heating season, simply multiply the collector-hours by two to obtain the total number of hours of motor time you will use, month by month.

If you want a quick estimate, use these multipliers times the *total* collector hours for the nine-month heating season:

Table 12 MULTIPLIERS FOR SOLAR FURNACE MOTOR OPERATION TIME

Percentage Supply From Solar Unit	Multiplier
50% or less	2.00
55%	1.95
60%	1.90
65%	1.85
70%	1.80
75%	1.75
80%	1.70
85%	1.65
90%	1.60

Note to Table 12: *Values ±10%, based upon calculations of motor operation time in selected cities. As percentage of heat supply from solar equipment approaches 100%, motor for distribution of heat to residence sees declining hours of operation, particularly in fall and in spring. It is assumed that flow rates of collection and distribution are equal; so daily distribution motor operation time cannot exceed daily collector motor operation time.*

Take that total and turn to Table 14, Cost to Run Solar Furnace Motors. Find the cost for electricity in your region and the size of the motors in your solar unit. Simply multiply your total motor hours by the number that appears where those columns intersect to obtain the cost to run the solar furnace for nine months. Divide by nine to find the average monthly cost.

Table 13 HOURS PER MONTH OF COLLECTOR MOTOR OPERATION

	Sep	Oct	Nov	Dec	Jan	Feb	Mar	Apr	May
ALABAMA									
Birmingham	244	234	182	136	138	152	207	248	293
Huntsville	242	239	180	128	134	147	204	243	281
Mobile	235	254	195	146	157	158	212	253	301
Montgomery	260	250	200	156	160	168	227	267	317
ARIZONA									
Flagstaff	298	274	218	214	187	215	270	298	346
Phoenix	334	307	267	236	248	244	314	346	404
Prescott	315	286	254	228	222	230	293	323	378
Tucson	335	317	280	258	255	266	317	350	399
Winslow	312	291	252	223	212	230	296	325	372
Yuma	351	330	285	262	258	266	337	365	419
ARKANSAS									
Fort Smith	261	230	174	147	146	156	202	234	268
Little Rock	265	251	181	142	143	158	213	243	291
Texarkana	272	268	194	163	157	169	218	246	297
CALIFORNIA									
Bakersfield	335	298	243	168	172	215	278	314	368
Bishop	321	244	177	164	200	198	259	276	351
Burbank	309	260	235	215	220	222	278	274	280
Eureka	195	164	127	108	120	138	180	209	247
Fresno	355	306	221	144	153	192	283	330	389
Long Beach	309	260	235	215	220	222	278	274	280
Los Angeles	295	263	249	220	224	217	273	264	292
Oakland	298	251	182	148	152	176	258	285	344
Red Bluff	341	277	199	154	156	186	246	302	366
Sacramento	347	283	197	122	134	169	255	300	367
San Diego	255	234	236	217	216	212	262	242	261
San Francisco	267	243	198	156	165	182	251	281	314
Santa Maria	290	256	221	198	200	221	259	267	281
COLORADO									
Alamosa	280	209	188	160	182	182	214	218	265
Colorado Springs	269	243	195	203	209	210	258	247	262
Denver	274	246	200	192	207	205	247	252	281
Grand Junction	291	255	198	168	169	182	243	265	314
Pueblo	290	265	225	211	224	217	261	271	299
CONNECTICUT									
Bridgeport	233	190	154	143	148	152	192	215	278
Hartford	220	193	137	136	141	166	206	223	267
New Haven	238	215	157	154	155	178	215	234	274

	Sep	Oct	Nov	Dec	Jan	Feb	Mar	Apr	May
DELAWARE									
Wilmington	232	212	159	142	135	168	210	231	272
DISTRICT OF COLUMBIA									
Washington	233	207	162	135	138	160	205	226	267
FLORIDA									
Appalachicola	236	263	216	175	193	195	233	274	328
Daytona Beach	204	206	200	191	202	200	246	274	290
Jacksonville	199	205	191	170	192	189	241	267	296
Key West	236	237	226	225	229	238	285	296	307
Miami Beach	216	215	212	209	222	227	266	275	280
Orlando	215	214	200	198	205	207	257	278	303
Pensacola	249	265	206	166	175	180	232	270	311
Tallahassee	226	249	208	178	181	187	241	271	309
Tampa	232	243	227	209	223	220	260	283	320
GEORGIA									
Athens	257	246	194	154	160	166	226	258	302
Atlanta	247	241	188	160	154	165	218	266	309
Augusta	260	254	192	161	164	167	229	273	305
Columbus	234	254	195	155	164	170	229	261	313
Macon	253	236	202	168	177	178	235	279	321
Rome	234	229	182	135	141	154	204	242	289
Savannah	212	216	197	167	175	173	229	274	307
IDAHO									
Boise	311	232	143	104	116	144	218	274	322
Idaho Falls	283	206	122	86	98	123	191	242	279
Lewiston	268	204	94	51	81	113	184	252	288
Pocatello	296	230	145	108	111	143	211	255	300
ILLINOIS									
Cairo	279	254	181	145	124	160	218	254	298
Chicago	247	216	136	118	126	142	199	221	274
Moline	251	214	130	123	132	139	189	214	255
Peoria	259	222	149	122	134	149	198	229	273
Rockford	248	210	133	119	129	134	199	220	276
Springfield	266	225	152	122	127	149	193	224	282
INDIANA									
Evansville	274	236	156	120	123	145	199	237	294
Fort Wayne	242	210	120	102	113	136	191	217	281
Indianapolis	265	222	139	118	118	140	193	227	278
South Bend	240	204	112	99	108	122	192	216	285
IOWA									
Burlington	270	243	175	147	148	165	217	241	284
Des Moines	263	227	156	136	155	170	203	236	276

	Sep	Oct	Nov	Dec	Jan	Feb	Mar	Apr	May
Dubuque	229	190	130	113	141	154	192	232	271
Sioux City	270	236	160	146	164	177	216	254	300
Waterloo	248	213	141	126	151	162	210	233	282
KANSAS									
Concordia	249	245	189	172	180	172	214	243	281
Dodge City	290	266	218	198	205	191	249	268	305
Goodland	273	247	199	193	205	208	251	262	297
Topeka	263	229	173	149	159	160	193	215	260
Wichita	277	245	206	182	187	186	233	254	291
KENTUCKY									
Covington	254	209	139	114	124	138	192	222	275
Lexington	242	206	146	107	116	130	181	218	265
Louisville	256	219	148	114	115	135	188	221	283
LOUISIANA									
Alexandria	267	266	200	158	151	162	208	260	289
Baton Rouge	256	256	194	151	157	156	211	237	285
Lake Charles	260	266	198	152	155	162	204	240	292
New Orleans	241	260	200	157	160	158	213	247	292
Shreveport	289	273	208	177	151	172	214	240	298
MAINE									
Caribou	159	142	79	74	88	116	151	180	210
Portland	229	202	146	148	155	174	213	226	268
MARYLAND									
Baltimore	238	212	164	145	148	170	211	229	270
Frederick	232	209	151	132	127	154	192	206	270
MASSACHUSETTS									
Boston	232	207	152	148	148	168	212	222	263
Nantucket	242	208	149	129	128	156	214	227	278
Pittsfield	188	172	111	87	119	148	188	205	246
Worcester	218	186	130	125	126	157	199	212	258
MICHIGAN									
Alpena	204	159	70	67	86	124	198	228	261
Detroit	226	189	98	89	90	128	180	212	263
Escanaba	198	162	90	94	112	148	204	226	266
Flint	218	172	88	76	93	145	188	213	264
Grand Rapids	231	188	92	70	74	117	178	218	277
Lansing	228	182	87	73	84	119	175	215	272
Marquette	186	142	68	66	78	113	172	207	248
Muskegon	222	165	88	78	84	115	188	221	268
Sault Ste. Marie	165	133	61	62	83	123	187	217	252
MINNESOTA									
Duluth	203	166	100	107	125	163	221	235	268

	Sep	*Oct*	*Nov*	*Dec*	*Jan*	*Feb*	*Mar*	*Apr*	*May*
International Falls	205	161	94	105	121	148	198	234	275
Minneapolis	237	193	115	112	140	166	200	231	272
Rochester	238	209	122	113	144	159	199	230	279
Saint Cloud	223	187	108	113	137	157	195	235	274
MISSISSIPPI									
Jackson	235	223	185	150	130	147	199	244	280
Meridian	253	237	186	147	156	155	215	253	295
Vicksburg	254	244	183	140	136	141	199	232	284
MISSOURI									
Columbia	262	225	166	138	147	164	207	232	281
Kansas City	266	235	178	151	154	170	211	235	278
St. Joseph	260	224	168	144	154	165	211	231	274
St. Louis	256	223	166	125	137	152	202	235	283
Springfield	269	233	183	140	145	164	213	238	278
MONTANA									
Billings	258	213	136	129	140	154	208	236	283
Glasgow	246	203	133	120	136	166	221	254	297
Great Falls	256	206	132	133	154	176	245	261	299
Havre	260	202	132	122	136	174	234	268	311
Helena	258	202	137	121	138	168	215	241	292
Kalispell	231	169	78	52	76	115	180	234	273
Miles City	249	214	147	135	144	176	224	257	299
Missoula	246	178	90	66	85	109	167	209	261
NEBRASKA									
Grand Island	266	232	175	157	178	185	229	247	292
Lincoln	266	237	174	160	173	172	213	244	287
Norfolk	255	221	162	142	167	178	221	244	289
North Platte	264	242	184	169	181	179	221	246	282
Omaha	270	248	166	145	172	188	222	259	305
Scottsbluff	263	235	177	173	185	196	240	248	280
NEVADA									
Elko	311	249	178	149	166	179	229	251	301
Ely	303	255	204	187	186	197	262	260	300
Las Vegas	345	301	258	250	239	251	314	336	386
Reno	336	273	212	170	185	199	267	306	354
Winnemucca	316	242	177	139	142	155	207	255	312
NEW HAMPSHIRE									
Concord	214	179	122	126	136	153	192	196	229
NEW JERSEY									
Atlantic City	239	218	177	153	151	173	210	233	273
Newark	240	211	157	143	145	167	210	235	278
Trenton	239	208	160	142	145	168	203	235	277

	Sep	Oct	Nov	Dec	Jan	Feb	Mar	Apr	May
NEW MEXICO									
Albuquerque	299	279	245	219	221	218	273	299	343
Raton	269	244	214	185	215	219	259	256	285
Roswell	266	266	242	216	218	223	286	306	330
Silver City	316	286	254	233	228	247	303	320	348
NEW YORK									
Albany	224	192	115	112	125	151	194	213	266
Binghampton	184	158	92	79	94	119	151	170	226
Buffalo	239	183	97	84	110	125	180	212	274
New York City	235	213	169	155	154	171	213	237	268
Rochester	224	173	97	86	93	123	172	209	274
Schenectady	218	186	114	106	125	150	196	213	259
Syracuse	211	163	81	74	87	115	165	197	261
NORTH CAROLINA									
Asheville	235	222	179	146	146	161	211	247	289
Charlotte	247	243	198	167	165	177	230	267	313
Greensboro	243	236	190	163	157	171	217	231	298
Raleigh	224	215	184	156	154	168	220	255	290
Wilmington	237	238	206	178	179	180	237	279	314
Winston-Salem	235	186	179	154	148	159	207	244	283
NORTH DAKOTA									
Bismarck	243	198	130	125	141	170	205	236	279
Devil's Lake	230	198	123	124	150	177	220	250	291
Fargo	222	187	112	114	132	170	210	232	283
Williston	247	206	131	129	141	168	215	260	305
OHIO									
Akron	233	190	98	72	86	108	166	207	279
Cincinnati	253	205	138	118	115	137	186	222	273
Cleveland	235	187	99	77	79	111	167	209	274
Columbus	250	210	131	101	112	132	177	215	270
Dayton	268	229	152	124	114	136	195	222	281
Mansfield	240	194	113	86	98	123	177	207	278
Sandusky	248	201	111	91	100	128	183	229	285
Toledo	241	196	106	92	93	120	170	203	263
Youngstown	229	180	95	72	83	117	166	200	269
OKLAHOMA									
Oklahoma City	282	243	201	175	175	182	235	253	290
Tulsa	281	241	207	172	152	164	200	213	244
OREGON									
Astoria	170	139	77	59	76	93	147	195	223
Eugene	241	144	81	55	69	97	147	202	229
Medford	248	173	112	71	82	116	177	208	249

Total (handwritten)

	Sep	Oct	Nov	Dec	Jan	Feb	Mar	Apr	May
Pendleton	280	204	108	56	98	111	184	248	329
Portland	218	134	87	65	77	97	142	203	246
Roseburg	255	146	81	50	69	96	148	205	257
Salem	226	144	81	55	74	97	147	203	230
PENNSYLVANIA									
Allentown	233	204	146	132	130	161	207	227	274
Erie	229	169	86	71	82	116	170	196	262
Harrisburg	233	200	140	131	132	160	203	230	277
Philadelphia *1742*	225	205	158	142	142	166	203	231	270
Pittsburgh	234	180	114	76	89	114	163	200	239
Reading *1646*	219	198	144	127	133	151	195	220	259
Scranton	213	183	120	105	108	138	178	199	251
Williamsport	206	173	104	89	101	126	177	196	242
RHODE ISLAND									
Providence	226	207	153	143	145	168	211	221	271
SOUTH CAROLINA									
Charleston	244	239	210	187	188	189	243	284	323
Columbia	243	242	202	166	173	183	233	274	312
Florence	234	236	197	154	163	172	226	262	293
Greenville	239	232	192	157	166	176	227	274	307
Spartanburg	231	225	183	146	150	163	207	239	277
SOUTH DAKOTA									
Huron	260	212	142	134	153	177	213	250	295
Rapid City	266	228	164	144	164	182	222	245	278
Sioux Falls	253	223	154	138	158	171	213	238	288
TENNESSEE									
Bristol	220	206	150	114	120	128	181	229	259
Chattanooga	247	220	169	128	126	146	187	239	290
Knoxville	237	213	157	120	124	144	189	237	281
Memphis	261	243	180	139	135	152	204	244	296
Nashville	250	224	168	126	123	142	196	241	285
TEXAS									
Abilene	275	245	223	201	190	199	250	259	290
Amarillo	288	260	229	205	207	199	258	276	305
Austin	261	242	180	160	148	152	207	221	266
Brownsville	246	252	165	151	147	152	187	210	272
Corpus Christi	276	264	194	164	160	165	212	237	295
Dallas	274	240	191	163	155	159	220	238	279
El Paso	300	287	257	236	234	236	299	329	373
Fort Worth	274	240	191	163	155	159	220	238	279
Galveston	257	264	199	151	151	149	203	230	288
Houston	238	239	181	146	144	141	193	212	266

	Sep	Oct	Nov	Dec	Jan	Feb	Mar	Apr	May
Laredo	274	250	207	177	180	188	241	250	273
Lubbock	286	271	241	215	226	222	278	293	332
Midland	298	276	246	218	233	217	278	292	337
Port Arthur	252	256	191	148	153	149	209	235	292
San Angelo	271	252	226	208	205	208	267	256	310
San Antonio	261	241	183	160	148	153	213	224	258
Victoria	256	263	193	156	163	157	204	243	290
Waco	268	248	190	165	163	164	226	245	278
Wichita Falls	275	243	200	175	179	179	237	254	284
UTAH									
Salt Lake City	306	249	171	135	137	155	227	269	329
VERMONT									
Burlington	199	152	77	80	103	127	184	185	244
VIRGINIA									
Lynchburg	235	217	177	158	153	169	216	243	288
Norfolk	237	220	182	161	156	174	223	257	304
Richmond	230	211	176	152	144	166	211	248	280
Roanoke	224	209	150	123	123	125	185	236	263
WASHINGTON									
Olympia	204	125	82	66	75	98	158	200	252
Seattle	197	122	77	62	74	99	154	201	247
Spokane	264	177	86	57	78	120	197	262	308
Tacoma	197	122	77	62	74	99	154	201	247
Walla Walla	280	198	92	51	72	106	194	262	317
Yakima	280	207	107	77	94	142	228	286	336
WEST VIRGINIA									
Charleston	216	188	116	95	100	115	162	190	238
Elkins	211	186	131	103	110	119	158	198	227
Huntington	224	195	122	98	103	121	166	198	256
Parkersburg	230	189	117	93	91	111	155	200	252
WISCONSIN									
Green Bay	213	176	110	106	121	148	194	210	251
La Crosse	234	199	125	108	141	150	195	234	270
Madison	230	198	116	108	126	147	196	214	258
Milwaukee	230	198	116	108	126	147	196	214	258
WYOMING									
Casper	263	217	164	157	180	195	251	261	278
Cheyenne	265	242	188	170	191	197	243	237	259
Lander	280	233	186	185	200	208	260	264	301
Sheridan	266	221	153	145	160	179	226	245	286
ALBERTA									
Banff	171	130	82	42	59	93	126	160	200

	Sep	Oct	Nov	Dec	Jan	Feb	Mar	Apr	May
Beaverlodge	176	139	83	62	75	106	162	214	270
Brooks	198	173	111	80	89	117	154	207	268
Calgary	188	166	116	94	99	121	156	196	237
Edmonton	185	161	105	80	91	113	176	224	272
Edson	162	150	79	68	85	115	152	190	234
Fairview	167	131	71	51	64	98	164	215	261
Lacombe	183	156	99	78	86	114	165	202	246
Lethbridge	215	178	114	96	107	129	171	206	260
Medicine Hat	188	165	105	86	91	118	149	199	256
Olds	186	165	106	72	89	119	157	189	231
Vauxhall	205	171	119	92	99	122	163	198	265
BRITISH COLUMBIA									
Campbell River	163	97	54	28	48	88	125	157	253
Dawson Creek	172	137	83	60	75	106	161	212	275
Fort St. John	173	137	82	59	73	105	161	213	278
Hope	148	91	41	33	43	68	95	115	166
Kamloops	190	113	56	32	49	84	150	166	227
Kitimat	122	63	36	30	47	66	124	156	213
Nanaimo	190	118	61	46	54	91	128	164	237
Nelson	190	104	54	25	34	82	139	153	243
New Westminster	182	118	69	46	48	91	132	164	246
Penticton	205	128	63	38	46	86	150	193	251
Port Alberni	190	118	61	46	54	91	128	164	237
Powell River	171	111	60	48	40	91	128	161	248
Prince George	157	105	59	39	54	88	139	185	257
Prince Rupert	96	53	41	23	42	60	88	118	158
Quesnel	156	106	60	40	55	88	139	184	255
Summerland	205	128	61	38	46	83	150	189	246
Terrace	122	63	36	30	47	66	124	156	213
Trail	190	104	39	13	24	71	128	140	213
Vancouver	182	118	69	46	48	91	132	164	246
Victoria	208	138	80	61	70	97	150	197	275
MANITOBA									
Brandon	188	160	84	85	106	127	164	200	239
Dauphin	175	155	88	92	116	138	176	217	264
Morden	191	169	89	91	107	136	160	200	238
Rivers	193	171	94	95	116	141	174	210	261
The Pas	167	142	52	63	78	117	166	199	254
Winnipeg	183	158	81	86	112	139	170	209	246
NEW BRUNSWICK									
Campbellton	168	126	77	82	95	122	153	179	224
Chatham	181	144	91	96	110	121	148	180	211

	Sep	Oct	Nov	Dec	Jan	Feb	Mar	Apr	May
Fredericton	166	140	85	91	103	118	141	160	201
Moncton	166	141	87	90	103	120	135	168	212
Saint John	175	153	90	108	110	125	153	159	193
NEWFOUNDLAND									
Gander	145	112	62	60	73	85	102	116	155
St. John's	150	109	65	56	70	78	92	114	166
NOVA SCOTIA									
Annapolis Royal	177	141	74	52	62	86	139	156	206
Halifax	176	154	93	85	93	117	144	162	204
Kentville	175	140	76	56	73	96	129	156	205
Sydney	168	139	74	67	81	106	126	161	204
Truro	152	127	76	67	85	106	126	161	203
Yarmouth	168	146	85	58	71	94	136	173	229
ONTARIO									
Atikokan	187	103	63	69	102	155	200	195	211
Belleville	172	151	79	81	84	109	136	162	207
Brampton	192	161	93	83	98	115	143	178	244
Chatham	190	167	89	72	78	98	125	165	217
Cornwall	172	136	74	65	90	118	158	184	229
Guelph	171	143	74	69	80	99	131	168	227
Hamilton	171	142	74	74	93	120	140	171	245
Harrow	191	163	82	71	77	97	126	162	229
Kapuskasing	122	92	40	52	76	104	143	165	190
Kingston	188	156	90	80	98	119	145	175	252
Lindsay	173	140	72	59	77	106	145	169	211
London	177	153	73	61	69	96	128	170	233
Mount Forest	172	141	57	61	71	119	144	187	248
New Liskeard	134	97	46	50	72	103	153	168	176
North Bay	158	115	59	70	97	130	158	188	231
Ottawa	171	138	76	78	96	115	150	175	231
St. Catharines	186	139	76	50	58	78	124	162	212
Sarnia	185	169	71	58	78	106	134	189	237
Sault Ste. Marie	159	121	62	66	74	114	160	186	246
Thunder Bay	173	116	80	92	116	149	185	205	235
Toronto	197	153	82	77	87	110	145	179	221
Woodstock	191	157	75	64	70	92	131	172	217
PRINCE EDWARD ISLAND									
Charlottetown	180	133	72	59	83	105	137	156	199
QUEBEC									
Amos	138	91	45	69	80	114	153	185	218
Berthierville	187	150	76	85	98	118	169	193	232
Grandes Bergeronnes	167	127	73	74	95	104	158	187	203

	Sep	Oct	Nov	Dec	Jan	Feb	Mar	Apr	May
L'Assomption	176	143	79	79	97	119	154	183	234
Lennoxville	166	135	67	60	79	102	135	162	212
Maniwaki	159	124	65	69	94	123	153	199	245
Montreal	196	150	75	79	96	122	171	188	241
Normandin	144	109	63	74	96	115	156	181	213
Quebec	167	126	63	65	81	99	139	163	198
Rimouski	159	103	59	43	62	81	124	166	182
Ste. Agathe-des-Monts	169	135	60	81	96	125	151	197	248
St. Ambroise	141	92	54	72	98	115	154	201	214
Sept-Iles	166	125	88	97	103	135	163	222	232
Sherbrooke	174	140	71	65	83	103	141	171	218
Victoriaville	183	136	69	70	89	111	159	183	225
SASKATCHEWAN									
Estevan	224	192	116	108	124	136	195	214	282
Indian Head	210	165	97	84	105	129	177	206	263
Melfort	163	144	75	67	98	117	159	199	242
Moose Jaw	207	175	101	83	100	117	160	214	267
Outlook	180	156	93	78	102	118	169	202	277
Prince Albert	170	143	80	72	94	115	168	212	259
Regina	194	169	96	83	98	117	156	210	271
Saskatoon	207	175	98	84	99	129	192	225	279
Scott	181	150	84	62	85	113	151	209	273
Swift Current	193	167	106	85	95	115	149	207	264
Weyburn	222	155	113	87	108	123	179	213	286
Yorkton	187	118	75	89	102	127	174	213	270

Notes to Table of Hours per Month of Collector Operation: *(1) Assumes no more collector operation time per month than mean hours of bright sunshine. (2) Values for some U.S. cities obtained from "Mean Total Number of Hours of Sunshine,"* Climatic Atlas of the United States, *Environmental Data Service, U.S. Department of Commerce, June 1968. (3) Some values for U.S. cities obtained as product of appropriate value for latitude from Table 28 (Possible Hours of Sunshine at Various Latitudes) times the value obtained from "Mean Percentage of Possible Sunshine,"* Climatic Atlas of the United States, *Environmental Data Service, U.S. Department of Commerce, June 1968. (4) Values for Canadian cities obtained from "Daily Bright Sunshine, 1941-1970," by Yorke, B.J., and Kendall, G.R., Atmospheric Environment Service, Department of the Environment, Downsview, Canada, 1972.*

Table 14 MULTIPLIERS FOR COST TO RUN SOLAR FURNACE MOTORS

Cost Per kwh (in cents)	1 hp	3/4hp	1/2hp	1/3hp	1/4hp	1/5hp	1/6hp	1/10hp	1/20hp
1	0.011	0.008	0.005	0.004	0.003	0.002	0.002	0.001	0.001
1-1/2	0.016	0.012	0.008	0.005	0.004	0.003	0.003	0.002	0.001
2	0.021	0.016	0.011	0.007	0.005	0.004	0.004	0.002	0.001
2-1/2	0.027	0.020	0.013	0.009	0.007	0.005	0.004	0.003	0.001
3	0.032	0.024	0.016	0.011	0.008	0.006	0.005	0.003	0.002
3-1/2	0.037	0.028	0.019	0.012	0.009	0.007	0.006	0.004	0.002
4	0.043	0.032	0.021	0.014	0.011	0.009	0.007	0.004	0.002
4-1/2	0.048	0.036	0.024	0.016	0.012	0.010	0.008	0.005	0.002
5	0.053	0.040	0.027	0.018	0.013	0.011	0.009	0.005	0.003
5-1/2	0.059	0.044	0.029	0.020	0.015	0.012	0.010	0.006	0.003
6	0.064	0.048	0.032	0.021	0.016	0.013	0.011	0.006	0.003
6-1/2	0.069	0.052	0.035	0.023	0.017	0.014	0.012	0.007	0.003
7	0.075	0.056	0.037	0.025	0.019	0.015	0.012	0.007	0.004
7-1/2	0.080	0.060	0.040	0.027	0.020	0.016	0.013	0.008	0.004
8	0.085	0.064	0.043	0.028	0.021	0.017	0.014	0.008	0.004
8-1/2	0.091	0.068	0.045	0.030	0.023	0.018	0.015	0.009	0.005
9	0.096	0.072	0.048	0.032	0.024	0.019	0.016	0.010	0.005
9-1/2	0.101	0.076	0.051	0.034	0.025	0.020	0.017	0.010	0.005
10	0.107	0.080	0.053	0.036	0.027	0.021	0.018	0.011	0.005
10-1/2	0.112	0.084	0.056	0.037	0.028	0.022	0.019	0.011	0.006
11	0.117	0.088	0.059	0.039	0.029	0.023	0.020	0.012	0.006
11-1/2	0.122	0.092	0.061	0.041	0.031	0.024	0.020	0.012	0.006
12	0.128	0.096	0.064	0.043	0.032	0.026	0.021	0.013	0.006
12-1/2	0.133	0.100	0.067	0.044	0.033	0.027	0.022	0.013	0.007

Notes to Table of Multipliers for Cost to Run Solar Furnace Motors: *(1) Assumes motor rated at proper horsepower and run at rated power draw. (2) Assumes 70% efficiency of motor.*

SECTION FIVE

When the Manufacturer Isn't Talking: How to Determine the Btu Output, Inside Your Home, of Any Kind of Flat-Plate or Vertical-Vane Solar Collector

13 Outfoxing the Manufacturer

CALCULATING THE USEFUL B.T.U. DELIVERY FROM ANY SYSTEM, INSIDE THE HOUSE

In Chapter Fourteen, the useful Btu delivery from a properly insulated (R-30) collector with two glass covers and a horizontal reflector, with a properly insulated (R-30) storage system using pebbles and located outside the house, has been precalculated for you, so that you need only look up the output (per square foot of collector) for your town or city.

I confess it is difficult to conceive of a reason why you would want to purchase a system with less insulation, or one that would not have reflectors or two covers. But, in the unlikely event that you like throwing your money away, or your brother-in-law manufactures a poorly designed system and you are stuck with buying from him, you will want to know what the Btu delivery is inside your house.

Unfortunately, very few manufacturers will tell you what the useful Btu delivery is from the system they are selling to you. I know that it hard to believe, but it is true.

Wait until you go shopping. Since you don't want to buy a pig in a poke, the only thing you can do is be a little bit smarter than the manufacturer who is selling you the solar furnace.

That means you are stuck with performing the calculations in this chapter.

You will need to find out several things on your first visit to the dealer's showroom. (1) Is the system a high-temperature (operating at 175° F or higher most of the time), a mid-temperature (operating at 140° F most of the time) or a low-temperature (operating below 110° F most of the time) solar furnace? (2) How many covers does the collector have? (3) How much insulation is there on the collector sides and back? (4) How much insulation is there on the storage unit? (5) Where will the storage be located? (6) What are the horsepower ratings of the collector and the distribution motors?

Several worksheets have been provided for your convenience in making the calculations. Simple addition, subtraction, multiplication and division are all that are required, but a pocket calculator helps too.

Step One: Find your home town in Table 15, Normal Heating Season Degree Days, below. Copy the monthly numbers onto your worksheet. If your town or city is not listed, select the nearest town that approximates the winter temperatures in your town.

If you live at a higher elevation, add nine Degree Days to each monthly number for every *hundred* feet your town is higher than the listed elevation. Example: Suppose that you live in Mountaintown, which is 30 miles from Plainscity and 1,000 feet higher in elevation. The monthly numbers for Plainscity are: September—140, October—360, etc. You would add 90 to each monthly number (1,000 feet = 10 X 100 feet = 10 X 9 Degree Days = 90 Degree Days per month). Therefore, you would enter 230 under September (140 + 90), 450 under October (360 + 90), etc.

Step Two: Refer to the Tables of Percentage Efficiency of Flat-Plate Collectors (Tables 16-27). There are sections for low, middle and high temperature systems, for single-cover and double-cover collectors, and for collectors using or not using horizontal reflectors. Under each section, various insulations are listed. Find the column that most closely describes the insulation used on the sides and back of the collector you wish to evaluate. Now refer to the Degree Days you have entered on your worksheet. Taking month by month, find where the Degree Days you have listed fall in the second column of the efficiency chart. Enter the percentage listed under the collector being evaluated on that line under "Percentage Collector Efficiency" on your worksheet under that month. For example: Suppose that you are evaluating a single-cover low-temperature collector with no reflector. Select that table below (Table 16). Suppose that the Degree Days in your city are: September—230, October—450, etc. If the collector has 1 inch of fibrous glass insulation (R-3.33), referring to that column in the table would yield values to be entered on your worksheet on the "Percentage Collector Efficiency" line as follows: September—71%, October—62%, etc.

Step Three: In the tables for Various Storage Modes (Tables 28-36), find the appropriate storage description. Using the same procedure as in Step Two, above, enter the efficiency percentages on your worksheet, month by month, on the "Percentage Storage Efficiency" line.

Step Four: On your worksheet for each month, multiply the percentages of efficiency for collector and for storage by each other, entering the result on the "Percentage System Efficiency" line. Example: Under September, the Percentage Collector Efficiency is 71%, the Percentage Storage Efficiency is 73%. Multiplying the two, you obtain 52% after rounding off your answer (0.71 X 0.73 = 0.5183, rounding = 0.52 = 52%). This would be the *system* efficiency for September. Follow this procedure for each month.

Step Five: Find the appropriate table of "Btu's Transmitted" for the collector you are assessing (see Tables 37-44). Find your latitude and simply copy the numbers from that line onto the "Btu's Transmitted" line on your worksheet.

Step Six: From the "Table of Possible Hours of Sunshine" (Table 45), find your latitude and copy the numbers from that line onto your worksheet on the "Possible Sunshine Hours" line (September through May)—the next to last line.

Step Seven: Referring to the "Table of Mean Percentage of Possible Sunshine" (Table 46), find your city or the nearest city with approximately similar sun conditions (cloudiness). Copy those numbers onto the

"Mean Percentage of Possible Sunshine" (the seventh line) on your worksheet.

Step Eight: Multiply the Btu's Transmitted, month by month (from Step Five) by the *system* efficiency (Step Four), entering the results on the "Cloudless Day Output" line.

Step Nine: Multiply the Cloudless Day Output, month by month, by the appropriate "Mean Percentage of Possible Sunshine" (from Step Seven), entering the result on the "Average-Day Btu Output Inside the House" line.

Step Ten: Multiply the Average-Day Btu Output by the number of days in the month (printed on the worksheet for your convenience) and enter the result under "Average Monthly Btu Output Inside the House."

Step Eleven: Add all the Average Monthly Btu Outputs (from Step Ten) together and divide by 273, entering the result in the square labeled "Average Daily Btu Output."

Step Twelve: Multiply the monthly numbers on the "Possible Sunshine Hours" (Step Six) by the "Mean Percentage of Possible Sunshine" (Step Seven) and enter the results under "Possible Hours of Collector Operation."

Step Thirteen: You may now use the Hot-Water-Collector-Gram, the Collector-Payback-Gram and the Cost-of-Operation charts in Chapter Twelve to form your own evaluation of the collector you are assessing.

WORKSHEET FOR CALCULATING B.T.U. OUTPUT, INSIDE THE HOUSE, OF COLLECTOR - STORAGE SYSTEM

	SEP	OCT	NOV	DEC	JAN	FEB	MAR	APR	MAY
Step 1 TOTAL NORMAL DEGREE DAYS									
Step 2 PERCENTAGE *COLLECTOR* EFFICIENCY									
Step 3 PERCENTAGE *STORAGE* EFFICIENCY									
Step 5 B.T.U.'S TRANSMITTED THROUGH COVER(S)									
Step 4 PERCENTAGE *SYSTEM* EFFICIENCY									
Step 8 CLOUDLESS-DAY B.T.U. OUTPUT INSIDE HOUSE									
Step 7 MEAN PERCENTAGE OF POSSIBLE SUNSHINE									
Step 9 AVERAGE DAY B.T.U. OUTPUT INSIDE HOUSE									
DAYS OF MONTH	x 30	x 31	x 30	x 31	x 31	x 28	x 31	x 30	x 31
Step 10 AVERAGE MONTHLY B.T.U. OUTPUT INSIDE HOUSE									
Step 6 POSSIBLE SUNSHINE HOURS									
Step 12 POSSIBLE HOURS OF COLLECTOR OPERATION									

Step 11 AVERAGE DAILY B.T.U. OUTPUT _____

NOTE: Do the steps in order, as described in the text.

WORKSHEET FOR CALCULATING B.T.U. OUTPUT, INSIDE THE HOUSE, OF COLLECTOR - STORAGE SYSTEM

	SEP	OCT	NOV	DEC	JAN	FEB	MAR	APR	MAY
Step1 TOTAL NORMAL DEGREE DAYS									
Step 2 PERCENTAGE *COLLECTOR* EFFICIENCY									
Step 3 PERCENTAGE *STORAGE* EFFICIENCY									
Step 5 B.T.U.'S TRANSMITTED THROUGH COVER(S)									
Step 4 PERCENTAGE *SYSTEM* EFFICIENCY									
Step 8 CLOUDLESS-DAY B.T.U. OUTPUT INSIDE HOUSE									
Step 7 MEAN PERCENTAGE OF POSSIBLE SUNSHINE									
Step 9 AVERAGE DAY B.T.U. OUTPUT INSIDE HOUSE									
DAYS OF MONTH	x 30	x 31	x 30	x 31	x 31	x 28	x 31	x 30	x 31
Step 10 AVERAGE MONTHLY B.T.U. OUTPUT INSIDE HOUSE									
Step 6 POSSIBLE SUNSHINE HOURS									
Step 12 POSSIBLE HOURS OF COLLECTOR OPERATION									

Step 11 AVERAGE DAILY B.T.U. OUTPUT _____

NOTE: Do the steps in order, as described in the text.

WORKSHEET FOR CALCULATING B.T.U. OUTPUT, INSIDE THE HOUSE, OF COLLECTOR - STORAGE SYSTEM

	SEP	OCT	NOV	DEC	JAN	FEB	MAR	APR	MAY
Step1 TOTAL NORMAL DEGREE DAYS									
Step 2 PERCENTAGE *COLLECTOR* EFFICIENCY									
Step 3 PERCENTAGE *STORAGE* EFFICIENCY									
Step 5 B.T.U.'S TRANSMITTED THROUGH COVER(S)									
Step 4 PERCENTAGE *SYSTEM* EFFICIENCY									
Step 8 CLOUDLESS-DAY B.T.U. OUTPUT INSIDE HOUSE									
Step 7 MEAN PERCENTAGE OF POSSIBLE SUNSHINE									
Step 9 AVERAGE DAY B.T.U. OUTPUT INSIDE HOUSE									
DAYS OF MONTH	x 30	x 31	x 30	x 31	x 31	x 28	x 31	x 30	x 31
Step 10 AVERAGE MONTHLY B.T.U. OUTPUT INSIDE HOUSE									
Step 6 POSSIBLE SUNSHINE HOURS									
Step 12 POSSIBLE HOURS OF COLLECTOR OPERATION									

Step 11 AVERAGE DAILY B.T.U. OUTPUT _____

NOTE: Do the steps in order, as described in the text.

WORKSHEET FOR CALCULATING B.T.U. OUTPUT, INSIDE THE HOUSE, OF COLLECTOR - STORAGE SYSTEM

	SEP	OCT	NOV	DEC	JAN	FEB	MAR	APR	MAY
Step1 TOTAL NORMAL DEGREE DAYS									
Step 2 PERCENTAGE *COLLECTOR* EFFICIENCY									
Step 3 PERCENTAGE *STORAGE* EFFICIENCY									
Step 5 B.T.U.'S TRANSMITTED THROUGH COVER(S)									
Step 4 PERCENTAGE *SYSTEM* EFFICIENCY									
Step 8 CLOUDLESS-DAY B.T.U. OUTPUT INSIDE HOUSE									
Step 7 MEAN PERCENTAGE OF POSSIBLE SUNSHINE									
Step 9 AVERAGE DAY B.T.U. OUTPUT INSIDE HOUSE									
DAYS OF MONTH	x 30	x 31	x 30	x 31	x 31	x 28	x 31	x 30	x 31
Step 10 AVERAGE MONTHLY B.T.U. OUTPUT INSIDE HOUSE									
Step 6 POSSIBLE SUNSHINE HOURS									
Step 12 POSSIBLE HOURS OF COLLECTOR OPERATION									

Step 11 AVERAGE DAILY B.T.U. OUTPUT _____

NOTE: Do the steps in order, as described in the text.

WORKSHEET FOR CALCULATING B.T.U. OUTPUT, INSIDE THE HOUSE, OF COLLECTOR - STORAGE SYSTEM

	SEP	OCT	NOV	DEC	JAN	FEB	MAR	APR	MAY
Step1 TOTAL NORMAL DEGREE DAYS									
Step 2 PERCENTAGE *COLLECTOR* EFFICIENCY									
Step 3 PERCENTAGE *STORAGE* EFFICIENCY									
Step 5 B.T.U.'S TRANSMITTED THROUGH COVER(S)									
Step 4 PERCENTAGE *SYSTEM* EFFICIENCY									
Step 8 CLOUDLESS-DAY B.T.U. OUTPUT INSIDE HOUSE									
Step 7 MEAN PERCENTAGE OF POSSIBLE SUNSHINE									
Step 9 AVERAGE DAY B.T.U. OUTPUT INSIDE HOUSE									
DAYS OF MONTH	x 30	x 31	x 30	x 31	x 31	x 28	x 31	x 30	x 31
Step 10 AVERAGE MONTHLY B.T.U. OUTPUT INSIDE HOUSE									
Step 6 POSSIBLE SUNSHINE HOURS									
Step 12 POSSIBLE HOURS OF COLLECTOR OPERATION									

Step 11 AVERAGE DAILY B.T.U. OUTPUT _____

NOTE: Do the steps in order, as described in the text.

Table 15 NORMAL HEATING DEGREE DAYS, BASE 65° F, 273-DAY SEASON

	Sep	Oct	Nov	Dec	Jan	Feb	Mar	Apr	May	TOTAL
ALABAMA										
Birmingham	6	93	363	555	592	462	363	108	9	2551
Huntsville	12	127	426	663	694	557	434	138	19	3070
Mobile	0	22	213	357	415	300	211	42	0	1560
Montgomery	0	68	330	527	543	417	316	90	0	2291
ARIZONA										
Flagstaff	201	558	867	1073	1169	991	911	651	437	6858
Phoenix	0	22	234	415	474	328	217	75	0	1765
Prescott	27	245	579	797	865	711	605	360	158	4347
Tucson	0	25	231	406	471	344	242	75	6	1800
Winslow	6	245	711	1008	1054	770	601	291	96	4782
Yuma	0	0	148	319	363	228	130	29	0	1217
ARKANSAS										
Fort Smith	12	127	450	704	781	596	456	144	22	3292
Little Rock	9	127	465	716	756	577	434	120	9	3213
Texarkana	0	78	345	561	628	468	350	105	0	2535
CALIFORNIA										
Bakersfield	0	37	282	502	546	364	267	105	19	2122
Bishop	42	248	576	797	874	666	539	306	143	4191
Burbank	6	43	177	301	366	277	239	138	81	1628
Eureka	258	329	414	499	546	470	505	438	372	3831
Fresno	0	78	339	558	586	406	319	150	56	2492
Long Beach	12	40	156	288	375	297	267	168	90	1693
Los Angeles	42	78	180	291	372	302	288	219	158	1930
Oakland	45	127	309	481	527	400	353	255	180	2677
Red Bluff	0	53	318	555	605	428	341	168	47	2515
Sacramento	12	81	363	577	614	442	360	216	102	2767
San Diego	15	37	123	251	313	249	202	123	84	1397
San Francisco	60	143	306	462	508	395	363	279	214	2730
Santa Maria	96	146	270	391	459	370	363	282	233	2610
COLORADO										
Alamosa	279	639	1065	1420	1476	1162	1020	696	440	8197
Colorado Springs	132	456	825	1032	1128	938	893	582	319	6305
Denver	117	428	819	1035	1132	938	887	558	288	6202
Grand Junction	30	313	786	1113	1209	907	729	387	146	5620
Pueblo	54	326	750	986	1085	871	772	429	179	5452
CONNECTICUT										
Bridgeport	66	307	615	986	1079	966	853	510	208	5590
Hartford	99	372	711	1119	1209	1061	899	495	177	6142
New Haven	87	347	648	1011	1097	991	871	543	245	5840

	Sep	Oct	Nov	Dec	Jan	Feb	Mar	Apr	May	TOTAL
DELAWARE										
Wilmington	51	270	588	927	980	874	735	387	112	4924
DISTRICT OF COLUMBIA										
Washington	33	217	519	834	871	762	626	288	74	4224
FLORIDA										
Appalachicola	0	16	153	319	347	260	180	33	0	1308
Daytona Beach	0	0	75	211	248	190	140	15	0	879
Jacksonville	0	12	144	310	332	246	174	21	0	1239
Key West	0	0	0	28	40	31	9	0	0	108
Miami Beach	0	0	0	40	56	36	9	0	0	141
Orlando	0	0	72	198	220	165	105	6	0	766
Pensacola	0	19	195	353	400	277	183	36	0	1463
Tallahassee	0	28	198	360	375	286	202	36	0	1485
Tampa	0	0	60	171	202	148	102	0	0	683
GEORGIA										
Athens	12	115	405	632	642	529	431	141	22	2929
Atlanta	18	127	414	626	639	529	437	168	25	2983
Augusta	0	78	333	552	549	445	350	90	0	2397
Columbus	0	87	333	543	552	434	338	96	0	2383
Macon	0	71	297	502	505	403	295	63	0	2136
Rome	24	161	474	701	710	577	468	177	34	3326
Savannah	0	47	246	437	437	353	254	45	0	1819
IDAHO										
Boise	132	415	792	1017	1113	854	722	438	245	5728
Idaho Falls	282	648	1107	1432	1600	1291	1107	657	388	8512
Lewiston	123	403	756	933	1063	815	694	426	239	5452
Pocatello	172	493	900	1166	1324	1058	905	555	319	6892
ILLINOIS										
Cairo	36	164	513	791	856	680	539	195	47	3821
Chicago	81	326	753	1113	1209	1044	890	480	211	6107
Moline	99	335	774	1181	1314	1100	918	450	189	6360
Peoria	87	326	759	1113	1218	1025	849	426	183	5986
Rockford	114	400	837	1221	1333	1137	961	516	236	6755
Springfield	72	291	696	1023	1135	935	769	354	136	5411
INDIANA										
Evansville	66	220	606	896	955	767	620	237	68	4435
Fort Wayne	105	378	783	1135	1178	1028	890	471	189	6157
Indianapolis	90	316	723	1051	1113	949	809	432	177	5660
South Bend	111	372	777	1125	1221	1070	933	525	239	6373
IOWA										
Burlington	93	322	768	1135	1259	1042	859	426	177	6081
Des Moines	99	363	837	1231	1398	1165	967	489	211	6760

	Sep	Oct	Nov	Dec	Jan	Feb	Mar	Apr	May	TOTAL
Dubuque	156	450	906	1287	1420	1204	1026	546	260	7255
Sioux City	108	369	867	1240	1435	1198	989	483	214	6903
Waterloo	138	428	909	1296	1460	1221	1023	531	229	7235
KANSAS										
Concordia	57	276	705	1023	1163	935	781	372	149	5461
Dodge City	33	251	666	939	1051	840	719	354	124	4977
Goodland	81	381	810	1073	1166	955	884	507	236	6093
Topeka	57	270	672	980	1122	893	722	330	125	5170
Wichita	33	229	618	905	1023	804	645	270	87	4614
KENTUCKY										
Covington	75	291	669	983	1035	893	756	390	149	5241
Lexington	54	239	609	902	946	818	685	325	105	4683
Louisville	54	248	609	890	930	818	682	315	105	4651
LOUISIANA										
Alexandria	0	56	273	431	471	361	260	69	0	1921
Baton Rouge	0	31	216	369	409	294	208	33	0	1560
Lake Charles	0	19	210	341	381	274	195	39	0	1459
New Orleans	0	19	192	322	363	258	192	39	0	1385
Shreveport	0	47	297	477	522	426	304	81	0	2184
MAINE										
Caribou	336	682	1044	1535	1690	1470	1308	858	468	9391
Portland	195	508	807	1215	1339	1182	1042	675	372	7335
MARYLAND										
Baltimore	48	264	585	905	936	820	679	327	90	4654
Frederick	66	307	624	955	995	876	741	384	127	5075
MASSACHUSETTS										
Boston	60	316	603	983	1088	972	846	513	208	5589
Nantucket	93	332	573	896	992	941	896	621	384	5728
Pittsfield	219	524	813	1231	1339	1196	1063	660	326	7389
Worcester	147	450	774	1171	1271	1123	998	612	304	6851
MICHIGAN										
Alpena	273	580	912	1268	1404	1299	1218	777	446	8177
Detroit	87	360	738	1088	1181	1058	936	522	220	6190
Escanaba	243	539	924	1293	1445	1296	1203	777	456	8176
Flint	159	465	843	1212	1330	1198	1066	639	319	7231
Grand Rapids	135	434	804	1147	1259	1134	1011	579	279	6782
Lansing	138	431	813	1163	1262	1142	1011	579	273	6812
Marquette	240	527	936	1268	1411	1268	1187	771	468	8076
Muskegon	120	400	762	1088	1209	1100	995	594	310	6578
Sault Ste. Marie	279	580	951	1367	1525	1380	1277	810	477	8646
MINNESOTA										
Duluth	330	632	1131	1581	1745	1518	1355	840	490	9622

	Sep	Oct	Nov	Dec	Jan	Feb	Mar	Apr	May	TOTAL
International Falls	363	701	1236	1724	1919	1621	1414	828	443	10249
Minneapolis	189	505	1014	1454	1631	1380	1166	621	288	8248
Rochester	186	474	1005	1438	1593	1366	1150	630	301	8143
Saint Cloud	225	549	1065	1500	1702	1445	1221	666	326	8699
MISSISSIPPI										
Jackson	0	65	315	502	546	414	310	87	0	2239
Meridian	0	81	339	518	543	417	310	81	0	2289
Vicksburg	0	53	279	462	512	384	282	69	0	2041
MISSOURI										
Columbia	54	251	651	967	1076	874	716	324	121	5034
Kansas City	39	220	612	905	1032	818	682	294	109	4711
St. Joseph	60	285	708	1039	1172	949	769	348	133	5463
St. Louis	60	251	627	936	1026	848	704	312	121	4885
Springfield	45	223	600	877	973	781	660	291	105	4555
MONTANA										
Billings	186	487	897	1135	1296	1100	970	570	285	6926
Glasgow	270	608	1104	1466	1711	1439	1187	648	335	8768
Great Falls	258	543	921	1169	1349	1154	1063	642	384	7483
Havre	306	595	1065	1367	1584	1364	1181	657	338	8457
Helena	294	601	1002	1265	1438	1170	1042	651	381	7844
Kalispell	321	654	1020	1240	1401	1134	1029	639	397	7835
Miles City	174	502	972	1296	1504	1252	1057	579	276	7612
Missoula	303	651	1035	1287	1420	1120	970	621	391	7798
NEBRASKA										
Grand Island	108	381	834	1172	1314	1089	908	462	211	6479
Lincoln	75	301	726	1066	1237	1016	834	402	171	5828
Norfolk	111	397	873	1234	1414	1179	983	498	233	6922
North Platte	123	440	885	1166	1271	1039	930	519	248	6621
Omaha	105	357	828	1175	1355	1126	939	465	208	6558
Scottsbluff	138	459	876	1128	1238	1008	921	552	285	6598
NEVADA										
Elko	225	561	924	1197	1314	1036	911	621	409	7198
Ely	234	592	939	1184	1308	1075	977	672	456	7437
Las Vegas	0	78	387	617	688	487	335	111	6	2709
Reno	204	490	801	1026	1073	823	729	510	357	6013
Winnemucca	210	536	876	1091	1172	916	837	573	363	6574
NEW HAMPSHIRE										
Concord	177	505	822	1240	1358	1184	1032	636	298	7252
NEW JERSEY										
Atlantic City	39	251	549	880	936	848	741	420	133	4797
Newark	30	248	573	921	983	876	729	381	118	4859
Trenton	57	264	576	924	989	885	753	399	121	4968

	Sep	Oct	Nov	Dec	Jan	Feb	Mar	Apr	May	TOTAL
NEW MEXICO										
Albuquerque	12	229	642	868	930	703	595	288	81	4348
Raton	126	431	825	1048	1116	904	834	543	301	6128
Roswell	18	202	573	806	840	641	481	201	31	3793
Silver City	6	183	525	729	791	605	518	261	87	3705
NEW YORK										
Albany	138	440	777	1194	1311	1156	992	564	239	6811
Binghampton	141	406	732	1107	1190	1081	949	543	229	6378
Buffalo	141	440	777	1156	1256	1145	1039	645	329	6928
New York City	30	233	540	902	986	885	760	408	118	4862
Rochester	126	415	747	1125	1234	1123	1014	597	279	6660
Schenectady	123	422	756	1159	1283	1131	970	543	211	6598
Syracuse	132	415	744	1153	1271	1140	1004	570	248	6677
NORTH CAROLINA										
Asheville	48	245	555	775	784	683	592	273	87	4042
Charlotte	6	124	438	691	691	582	481	156	22	3191
Greensboro	33	192	513	778	784	672	552	234	47	3805
Raleigh	21	164	450	716	725	616	487	180	34	3393
Wilmington	0	74	291	521	546	462	357	96	0	2347
Winston-Salem	21	171	483	747	753	652	524	207	37	3595
NORTH DAKOTA										
Bismarck	222	577	1083	1463	1708	1442	1203	645	329	8672
Devils Lake	273	642	1191	1634	1872	1579	1345	753	381	9670
Fargo	219	574	1107	1569	1789	1520	1262	690	332	9062
Williston	261	601	1122	1513	1758	1473	1262	681	357	9028
OHIO										
Akron	96	381	726	1070	1138	1016	871	489	202	5989
Cincinnati	54	248	612	921	970	837	701	336	118	4797
Cleveland	105	384	738	1088	1159	1047	918	552	260	6251
Columbus	84	347	714	1039	1088	949	809	426	171	5627
Dayton	78	310	696	1045	1097	955	809	429	167	5586
Mansfield	114	397	768	1110	1169	1042	924	543	245	6312
Sandusky	66	313	684	1032	1107	991	868	495	198	5754
Toledo	117	406	792	1138	1200	1056	924	543	242	6418
Youngstown	120	412	771	1104	1169	1047	921	540	248	6332
OKLAHOMA										
Oklahoma City	15	164	498	766	868	664	527	189	34	3725
Tulsa	18	158	522	787	893	683	539	213	47	3860
OREGON										
Astoria	210	375	561	679	753	622	636	480	363	4679
Eugene	129	366	585	719	803	627	589	426	279	4523
Medford	78	372	678	871	918	697	642	432	242	4930

	Sep	*Oct*	*Nov*	*Dec*	*Jan*	*Feb*	*Mar*	*Apr*	*May*	*TOTAL*
Pendleton	111	350	711	884	1017	773	617	396	205	5064
Portland	114	335	597	735	825	644	586	396	245	4478
Roseburg	105	329	567	713	766	608	570	405	267	4330
Salem	111	338	594	729	822	647	611	417	273	4542
PENNSYLVANIA										
Allentown	90	353	693	1045	1116	1002	849	471	167	5786
Erie	102	391	714	1063	1169	1081	973	585	288	6366
Harrisburg	63	298	648	992	1045	907	766	396	124	5239
Philadelphia	60	291	621	964	1014	890	744	390	115	5089
Pittsburgh	105	375	726	1063	1119	1002	874	480	195	5939
Reading	54	257	597	939	1001	885	735	372	105	4945
Scranton	132	434	762	1104	1156	1028	893	498	195	6202
Williamsport	111	375	717	1073	1122	1002	856	468	177	5901
RHODE ISLAND										
Providence	96	372	660	1023	1110	988	868	534	236	5887
SOUTH CAROLINA										
Charleston	0	59	282	471	487	389	291	54	0	2033
Columbia	0	84	345	577	570	470	357	81	0	2484
Florence	0	78	315	552	552	459	347	84	0	2387
Greenville	0	112	387	636	648	535	434	120	12	2884
Spartanburg	15	130	417	667	663	560	453	144	25	3074
SOUTH DAKOTA										
Huron	165	508	1014	1432	1628	1355	1125	600	288	8115
Rapid City	165	481	897	1172	1333	1145	1051	615	326	7185
Sioux Falls	168	462	972	1361	1544	1285	1082	573	270	7717
TENNESSEE										
Bristol	51	236	573	828	828	700	598	261	68	4143
Chattanooga	18	143	468	698	722	577	453	150	25	3254
Knoxville	30	171	489	725	732	613	493	198	43	3494
Memphis	18	130	447	698	729	585	456	147	22	3232
Nashville	30	158	495	732	778	644	512	189	40	3578
TEXAS										
Abilene	0	99	366	586	642	470	347	114	0	2624
Amarillo	18	205	570	797	877	664	546	252	56	3985
Austin	0	31	225	388	468	325	223	51	0	1711
Brownsville	0	0	66	149	205	106	74	0	0	600
Corpus Christi	0	0	120	220	291	174	109	0	0	914
Dallas	0	62	321	524	601	440	319	90	6	2363
El Paso	0	84	414	648	685	445	319	105	0	2700
Fort Worth	0	65	324	536	614	448	319	99	0	2405
Galveston	0	0	138	270	350	258	189	30	0	1235
Houston	0	6	183	307	384	288	192	36	0	1396

	Sep	Oct	Nov	Dec	Jan	Feb	Mar	Apr	May	TOTAL
Laredo	0	0	105	217	267	134	74	0	0	797
Lubbock	18	174	513	744	800	613	484	201	31	3578
Midland	0	87	381	592	651	468	322	90	0	2591
Port Arthur	0	22	207	329	384	274	192	39	0	1447
San Angelo	0	68	318	536	567	412	288	66	0	2255
San Antonio	0	31	207	363	428	286	195	39	0	1549
Victoria	0	6	150	270	344	230	152	21	0	1173
Waco	0	43	270	456	536	389	270	66	0	2030
Wichita Falls	0	99	381	632	698	518	378	120	6	2832
UTAH										
Salt Lake City	81	419	849	1082	1172	910	763	459	233	5968
VERMONT										
Burlington	207	539	891	1349	1513	1333	1187	714	353	8086
VIRGINIA										
Lynchburg	51	223	540	822	849	731	605	267	78	4166
Norfolk	0	136	408	698	738	655	533	216	37	3421
Richmond	36	214	495	784	815	703	546	219	53	3865
Roanoke	51	229	549	825	834	722	614	261	65	4150
WASHINGTON										
Olympia	198	422	636	753	834	675	645	450	307	4920
Seattle	129	329	543	657	738	599	577	396 ·	242	4210
Spokane	168	493	879	1082	1231	980	834	531	288	6486
Tacoma	162	391	633	750	828	678	657	474	295	4868
Walla Walla	87	310	681	843	986	745	589	342	177	4760
Yakima	144	450	828	1039	1163	868	713	435	220	5860
WEST VIRGINIA										
Charleston	63	254	591	865	880	770	648	300	96	4467
Elkins	135	400	729	992	1008	896	791	444	198	5593
Huntington	63	257	585	856	880	764	636	294	99	4434
Parkersburg	60	264	606	905	942	826	691	339	115	4748
WISCONSIN										
Green Bay	174	484	924	1333	1494	1313	1141	654	335	7852
La Crosse	153	437	924	1339	1504	1277	1070	540	245	7489
Madison	174	474	930	1330	1473	1274	1113	618	310	7696
Milwaukee	174	471	876	1252	1376	1193	1054	642	372	7410
WYOMING										
Casper	192	524	942	1169	1290	1084	1020	657	381	7259
Cheyenne	210	543	924	1101	1228	1056	1011	672	381	7126
Lander	204	555	1020	1299	1417	1145	1017	654	381	7692
Sheridan	219	539	948	1200	1355	1154	1054	642	366	7477
ALBERTA										
Banff	501	791	1194	1510	1646	1266	1234	864	605	9611

	Sep	Oct	Nov	Dec	Jan	Feb	Mar	Apr	May	TOTAL
Beaverlodge	459	772	1272	1655	1854	1442	1345	849	493	10141
Brooks	354	670	1184	1525	1829	1383	1268	765	409	9387
Calgary	414	707	1131	1445	1634	1299	1262	810	502	9204
Edmonton	402	719	1215	1618	1845	1453	1327	774	415	9768
Edson	501	812	1272	1640	1804	1378	1287	843	546	10083
Fairview	462	806	1344	1761	1978	1546	1411	864	471	10643
Lacombe	432	753	1230	1631	1851	1459	1349	813	474	9992
Lethbridge	339	614	1035	1345	1566	1232	1169	711	434	8445
Medicine Hat	279	601	1110	1445	1699	1333	1203	669	347	8686
Olds	429	732	1167	1491	1711	1355	1308	822	508	9523
Vauxhall	327	636	1068	1423	1724	1291	1187	744	406	8806
BRITISH COLUMBIA										
Campbell River	252	524	726	877	949	745	769	588	384	5814
Dawson Creek	468	778	1323	1748	1953	1532	1392	852	512	10467
Fort St. John	453	791	1341	1764	1984	1526	1383	855	474	10571
Hope	150	440	732	936	1054	750	704	483	285	5534
Kamloops	168	530	867	1159	1349	958	791	468	211	6501
Kitimat	294	605	879	1094	1243	921	899	663	446	7044
Nanaimo	195	468	681	809	877	706	716	522	329	5303
Nelson	198	555	858	1088	1200	927	853	537	288	6504
New Westminster	174	425	666	831	902	700	685	477	301	5161
Penticton	198	536	825	1045	1187	907	818	519	276	6311
Port Alberni	207	481	732	890	989	795	760	558	347	5759
Powell River	159	419	627	772	846	680	685	495	295	4978
Prince George	459	763	1143	1448	1683	1238	1138	777	496	9145
Prince Rupert	342	539	714	859	921	787	815	654	487	6118
Quesnel	357	685	1050	1364	1575	1140	1008	669	397	8245
Summerland	159	521	852	1088	1221	924	815	504	254	6338
Terrace	306	614	903	1150	1302	955	893	672	412	7207
Trail	36	543	876	1122	1243	946	831	501	260	6358
Vancouver	207	443	651	794	868	689	694	525	53	4924
Victoria	219	415	600	732	797	633	642	489	347	4874
MANITOBA										
Brandon	360	710	1278	1792	2058	1694	1510	849	471	10722
Dauphin	366	685	1263	1798	2062	1705	1535	867	477	10758
Morden	285	620	1182	1696	1978	1616	1411	783	403	9974
Rivers	360	707	1287	1804	2058	1708	1507	834	453	10718
The Pas	432	794	1416	1996	2251	1848	1637	969	539	11882
Winnipeg	312	657	1230	1786	2043	1716	1476	810	431	10461
NEW BRUNSWICK										
Campbellton	303	642	966	1442	1628	1434	1259	858	508	9040
Chatham	267	605	921	1389	1544	1364	1215	834	493	8632

	Sep	Oct	Nov	Dec	Jan	Feb	Mar	Apr	May	TOTAL
Fredericton	225	567	903	1389	1538	1330	1135	750	409	8246
Moncton	261	574	867	1318	1466	1305	1181	822	487	8281
Saint John	243	518	792	809	1386	1212	1091	759	493	7303
NEWFOUNDLAND										
Gander	354	670	888	1215	1364	1240	1221	945	670	8567
St. John's	330	589	771	1045	1200	1114	1122	891	673	7735
NOVA SCOTIA										
Annapolis Royal	228	490	732	1101	1240	1134	1032	723	450	7130
Halifax	150	428	684	1054	1200	1089	1017	732	481	6835
Kentville	210	508	771	1156	1311	1190	1091	756	450	7443
Sydney	237	527	762	1107	1271	1201	1156	873	595	7729
Truro	246	539	780	1194	1339	1226	1122	783	496	7725
Yarmouth	240	474	696	1032	1172	1072	1011	741	505	6943
ONTARIO										
Atikokan	450	747	1257	1795	2055	1733	1473	894	564	10968
Belleville	138	471	810	1271	1442	1252	1079	636	322	7421
Brampton	168	481	819	1225	1367	1210	1110	654	347	7381
Chatham	48	363	729	1122	1240	1098	955	564	248	6367
Cornwall	171	499	852	1355	1541	1350	1156	648	304	7876
Guelph	180	502	849	1262	1407	1246	1113	669	366	7594
Hamilton	87	394	744	1101	1243	1103	980	588	291	6531
Harrow	30	353	738	1119	1243	1078	942	543	226	6272
Kapuskasing	414	725	1209	1814	2046	1730	1544	954	586	11022
Kingston	111	477	786	1203	1417	1310	1138	705	347	7494
Lindsay	201	530	885	1361	1516	1324	1163	684	375	8039
London	144	468	822	1221	1358	1207	1063	633	335	7251
Mount Forest	246	564	900	1358	1507	1364	1194	747	453	8333
New Liskeard	324	626	1065	1618	1879	1574	1411	870	487	9854
North Bay	261	586	957	1500	1708	1448	1259	762	403	8884
Ottawa	189	527	888	1420	1612	1369	1178	657	322	8162
St. Catharines	57	369	720	1085	1215	1078	961	579	295	6359
Sarnia	123	400	771	1119	1290	1151	1042	600	338	6834
Sault Ste. Marie	330	601	951	1414	1550	1397	1287	840	561	8931
Thunder Bay	378	685	1125	1624	1848	1579	1367	861	561	10028
Toronto	72	397	729	1122	1268	1114	989	582	285	6558
Woodstock	162	477	828	1218	1364	1207	1076	648	344	7324
PRINCE EDWARD ISLAND										
Charlottetown	204	512	789	1203	1389	1266	1175	831	508	7877
QUEBEC										
Amos	414	750	1173	1783	2018	1708	1516	957	561	10880
Berthierville	243	570	939	1500	1708	1481	1256	747	369	8813
Grandes Bergeronnes	405	713	1020	1525	1727	1467	1321	906	574	9656

L'Assomption	216	552	915	1485	1686	1456	1228	711	347	8596
Lennoxville	249	574	915	1438	1615	1403	1200	732	394	8520
Maniwaki	321	642	987	1566	1773	1518	1318	786	443	9354
Montreal	168	496	858	1373	1553	1355	1172	660	298	7933
Normandin	405	735	1131	1779	2027	1722	1504	966	555	10824
Quebec	228	577	945	1460	1615	1380	1197	768	391	8561
Rimouski	354	660	957	1432	1624	1406	1277	849	546	9105
Ste. Agathe-des-Monts	348	663	1050	1556	1724	1476	1299	822	446	9384
St. Ambroise	387	722	1110	1736	1978	1663	1423	909	530	10458
Sept-Iles	471	800	1107	1603	1798	1562	1398	1008	707	10454
Sherbrooke	219	543	885	1395	1559	1355	1156	711	353	8176
Victoriaville	267	592	951	1451	1637	1428	1246	747	397	8716
SASKATCHEWAN										
Estevan	315	657	1191	1665	1907	1560	1376	777	415	9863
Indian Head	360	710	1260	1742	1972	1632	1466	822	440	10404
Melfort	420	781	1359	1838	2086	1733	1569	909	465	11160
Moose Jaw	312	648	1164	1603	1879	1492	1376	750	397	9621
Outlook	336	691	1230	1683	1938	1582	1411	771	391	10033
Prince Albert	438	803	1371	1938	2198	1775	1606	897	493	11519
Regina	366	729	1272	1742	1987	1646	1488	810	434	10474
Saskatoon	381	719	1278	1783	1987	1649	1466	798	425	10486
Scott	435	794	1344	1820	2065	1683	1528	843	474	10986
Swift Current	354	682	1185	1544	1783	1467	1324	777	434	9550
Weyburn	324	676	1185	1621	1891	1543	1386	786	409	9821
Yorkton	390	738	1314	1826	2083	1714	1559	876	471	10971

Notes to Table of Total Normal Heating Degree Days (Base 65°F): *(1) Values for U.S. Cities obtained from "Normal Total Heating Degree Days (Base 65°)"*, Climatic Atlas of the United States, *Environmental Data Service, U.S. Department of Commerce, June 1968. (2) Values for Canadian Cities obtained by subtracting "Mean Daily Temperature (Deg. F)"*, Canadian Normals, Temperature, 1941-1970, *Volume 1, Atmospheric Environment Service, Department of the Environment of Canada, Downsview, Ontario, 1973, from the base temperature of 65°F and multiplying times the number of days in the respective month.*

Table 16 AVERAGE PERCENTAGE EFFICIENCY OF SINGLE-COVER LOW-TEMPERATURE COLLECTOR WITHOUT REFLECTORS

Outside Temp. (F)	Degree Days Per Month	Glass-Fiber Batt Insulation				Urethane Foam		
		R-1.67, 1/2"	R-3.33, 1"	R-6.67, 2"	R-8.33, 2-1/2"	R-10, 3"	R-15, 2"	R-30, 4"
65	0- 149	71	76	79	80	80	81	81
60	150- 299	65	71	75	75	76	77	77
55	300- 449	59	67	71	71	72	73	74
50	450- 599	53	62	66	67	68	69	70
45	600- 749	47	57	62	63	64	65	66
40	750- 899	41	52	57	59	60	61	62
35	900-1049	35	48	54	55	56	57	59
30	1050-1199	30	43	50	51	52	53	55
25	1200-1349	24	38	45	47	48	49	51
20	1350-1499	18	33	41	43	44	46	47
15	1500-1649	12	29	37	39	40	42	44
10	1650-1799	06	24	33	35	36	38	40
5	1800-1949	0	19	29	30	32	34	36
0	1950-2099		14	24	26	28	30	32
5	2100-2249		10	20	22	24	26	29
10	2250-2399		05	16	18	20	22	25
15	2400-2549		0	12	14	16	18	21

Table 17 AVERAGE PERCENTAGE EFFICIENCY OF SINGLE-COVER LOW-TEMPERATURE COLLECTOR WITH REFLECTORS

Outside Temp. (F)	Degree Days Per Month	Glass-Fiber Batt Insulation				Urethane Foam		
		R-1.67, 1/2"	R-3.33, 1"	R-6.67, 2"	R-8.33, 2-1/2"	R-10, 3"	R-15, 2"	R-30, 4"
65	0- 149	75	80	82	83	83	84	84
60	150- 299	70	76	79	79	80	80	81
55	300- 449	66	72	75	76	76	77	78
50	450- 599	61	68	72	73	73	74	75
45	600- 749	56	64	68	69	70	71	72
40	750- 899	51	60	64	66	63	64	65
35	900-1049	46	56	61	62	60	61	62
30	1050-1199	41	52	58	59	60	61	62
25	1200-1349	36	48	54	55	56	58	59
20	1350-1499	31	44	51	52	53	54	56
15	1500-1649	26	40	47	49	50	51	53
10	1650-1799	21	36	44	45	46	48	50
5	1800-1949	16	32	40	42	43	45	47
0	1950-2099	11	28	37	38	39	41	43
- 5	2100-2249	07	24	33	35	36	38	40
- 10	2250-2399	02	20	30	31	33	35	37
- 15	2400-2549	0	16	26	28	29	32	34

Notes to Tables 16 and 17: *(1) Glass: single -1.13 Btu/ft^2·hr·°F, double—0.45 Btu/ft^2·hr·°F; fibrous glass—0.3 Btu/ft^2·hr·°F/in.; urethane foam—0.133 Btu/ft^2·hr·°F/in. (2) Efficiency defined as a ratio between energy actually transmitted through covers to energy transferred to storage, expressed as a percentage. (3) Low-temperature collector: Mean Absorber Temperature (T_i + T_o/2) = 90° F. (4) Average (daily, over 273-day heating season) energy transmitted through covers, assuming full-day random distribution of cloudiness; single cover, no reflector—155 Btu/ft^2·hr; single cover with reflector at horizontal—185 Btu/ft^2·hr. (5) 15% side area per ft^2*

Table 18 AVERAGE PERCENTAGE EFFICIENCY OF SINGLE-COVER MID-TEMPERATURE COLLECTOR WITHOUT REFLECTORS (FOR DOMESTIC HOT WATER)

Outside Temp. (F)	Degree Days Per Month	Glass-Fiber Batt Insulation					Urethane Foam	
		R-1.67, 1/2"	R-3.33, 1"	R-6.67, 2"	R-8.33, 2-1/2"	R-10, 3"	R-15, 2"	R-30, 4"
65	0- 149	12	29	37	39	40	42	43
60	150- 299	06	24	33	35	36	38	40
55	300- 449	0	19	29	30	32	34	36
50	450- 599		14	24	26	28	30	32
45	600- 749		10	20	22	24	26	28
40	750- 899		05	16	18	20	22	25
35	900-1049		0	12	14	16	18	21
30	1050-1199			08	10	12	14	17
25	1200-1349			03	06	08	10	13
20	1350-1499			0	02	04	07	10
15	1500-1649				0	0	03	06
10	1650-1799						0	02
5	1800-1949							0

Table 19 AVERAGE PERCENTAGE EFFICIENCY OF SINGLE-COVER MID-TEMPERATURE COLLECTOR WITH REFLECTORS (FOR DOMESTIC HOT WATER)

Outside Temp. (F)	Degree Days Per Month	Glass-Fiber Batt Insulation					Urethane Foam	
		R-1.67, 1/2"	R-3.33, 1"	R-6.67, 2"	R-8.33, 2-1/2"	R-10, 3"	R-15, 2"	R-30, 4"
65	0- 149	26	40	47	49	50	51	53
60	150- 299	21	36	44	45	46	48	49
55	300- 449	16	32	40	42	43	45	46
50	450- 599	11	28	37	38	39	41	43
45	600- 749	07	24	33	35	36	38	40
40	750- 899	02	20	30	31	33	35	37
35	900-1049	0	16	26	28	29	32	34
30	1050-1199		12	23	25	26	28	31
25	1200-1349		08	19	21	23	25	27
20	1350-1499		04	16	18	19	22	24
15	1500-1649		0	12	14	16	18	21
10	1650-1799			08	11	13	15	18
5	1800-1949			05	07	09	12	15
0	1950-2099			01	04	06	09	12
− 5	2100-2249			0	01	03	05	08
− 10	2250-2399				0	0	02	05
− 15	2400-2549						0	02

Notes to Tables 18 and 19: (1) Glass: single—1.13 $Btu/ft^2 \cdot hr \cdot °F$, double—0.45 $Btu/ft^2 \cdot hr \cdot °F$; fibrous glass— 0.3 $Btu/ft^2 \cdot hr \cdot °F/in.$; urethane foam—0.133 $Btu/ft^2 \cdot hr \cdot °F/in.$ (2) Efficiency defined as ratio between energy actually transmitted through covers to energy transferred to storage, expressed as a percentage. (3) Mid-temperature collector: Mean Absorber Temperature $(T_i + T_o/2) = 140°F$. (4) Average (daily, over 273-day heating season) energy transmitted through covers, assuming full-day random distribution of cloudiness: single cover, no reflector—155 $Btu/ft^2 \cdot hr$; single cover with reflector at horizontal—185 $Btu/ft^2 \cdot hr$. (5) 15% side area per ft^2.

Table 20 AVERAGE PERCENTAGE EFFICIENCY OF SINGLE-COVER HIGH-TEMPERATURE COLLECTOR WITHOUT REFLECTORS

Outside Temp. (F)	Degree Days Per Month	Glass-Fiber Batt Insulation					Urethane Foam	
		R-1.67, 1/2"	R-3.33, 1"	R-6.67, 2"	R-8.33, 2-1/2"	R-10, 3"	R-15, 2"	R-30, 4"
65	0- 149	0	12	23	25	26	28	31
60	150- 299		08	19	21	23	25	28
55	300- 449		04	16	18	19	22	25
50	450- 599		0	12	14	16	18	21
45	600- 749			08	11	13	15	18
40	750- 899			05	07	09	12	15
35	900-1049			01	04	06	09	12
30	1050-1199			0	01	02	05	09
25	1200-1349				0	0	02	06
20	1350-1499						0	03
15	1500-1649							0

Table 21 AVERAGE PERCENTAGE EFFICIENCY OF SINGLE-COVER HIGH-TEMPERATURE COLLECTOR WITH REFLECTORS

Outside Temp. (F)	Degree Days Per Month	Glass-Fiber Batt Insulation					Urethane Foam	
		R-1.67, 1/2"	R-3.33, 1"	R-6.67, 2"	R-8.33, 2-1/2"	R-10, 3"	R-15, 2"	R-30, 4"
65	0- 149	0	0	08	10	12	14	17
60	150- 299			03	06	08	10	14
55	300- 449			0	02	04	07	10
50	450- 599				0	0	03	06
45	600- 749						0	02
40	750- 899							0

Notes to Tables 20 and 21: *(1) Glass: single pane—1.13 Btu/ft^2·hr·°F; fibrous glass batt—0.3 Btu/ft^2·hr·°F/ in.; urethane foam—0.133 Btu/ft^2·hr·°F/in. (2) Efficiency defined as ratio between energy actually transmitted through covers to energy transferred to storage, expressed as a percentage. (3) High-temperature collector:Mean Absorber Temperature (T_i + T_o/2) = 175°F. (4) Average (daily, over 273-day heating season) energy transmitted through covers, assuming full-day random distribution of cloudiness: single cover, with no reflector—155 Btu/ft^2·hr; single cover with reflector at horizontal—185 Btu/ft^2·hr. (5) 15% side area assumed per ft^2.*

Table 22 AVERAGE PERCENTAGE EFFICIENCY OF DOUBLE-COVER LOW-TEMPERATURE COLLECTOR WITHOUT REFLECTORS

Outside Temp. (F)	Degree Days Per Month	Glass-Fiber Batt Insulation					Urethane Foam	
		R-1.67, 1/2"	R-3.33, 1"	R-6.67, 2"	R-8.33, 2-1/2"	R-10, 3"	R-15, 2"	R-30, 4"
65	0- 149	78	85	88	89	89	90	91
60	150- 299	74	82	86	86	87	88	89
55	300- 449	69	79	83	84	85	86	87
50	450- 599	65	76	81	82	83	84	85
45	600- 749	61	72	78	80	80	82	83
40	750- 899	56	69	76	77	78	80	81
35	900-1049	52	66	74	75	76	78	80
30	1050-1199	47	63	71	73	74	76	78
25	1200-1349	43	60	69	71	72	74	76
20	1350-1499	39	57	66	68	70	72	74
15	1500-1649	34	54	64	66	67	70	72
10	1650-1799	30	51	62	64	65	68	70
5	1800-1949	25	48	59	62	63	66	68
0	1950-2099	21	45	57	59	61	64	67
− 5	2100-2249	17	42	55	57	59	62	65
− 10	2250-2399	12	39	52	55	57	60	63
− 15	2400-2549	08	36	50	53	54	57	61

Table 23 AVERAGE PERCENTAGE EFFICIENCY OF DOUBLE-COVER LOW-TEMPERATURE COLLECTOR WITH REFLECTORS

Outside Temp. (F)	Degree Days Per Month	Glass-Fiber Batt Insulation					Urethane Foam	
		R-1.67, 1/2"	R-3.33, 1"	R-6.67, 2"	R-8.33, 2-1/2"	R-10, 3"	R-15, 2"	R-30, 4"
65	0- 149	82	87	90	91	91	92	93
60	150- 299	78	85	88	89	89	90	91
55	300- 449	74	82	86	87	87	88	89
50	450- 599	71	79	84	85	85	86	88
45	600- 749	67	77	82	83	83	85	86
40	750- 899	63	74	80	81	82	83	84
35	900-1049	60	72	78	79	80	81	83
30	1050-1199	56	69	76	77	78	80	81
25	1200-1349	52	67	74	75	76	78	80
20	1350-1499	49	64	72	73	74	76	78
15	1500-1649	45	62	70	72	73	75	77
10	1650-1799	41	59	68	70	71	73	75
5	1800-1949	37	56	66	68	69	71	73
0	1950-2099	34	54	64	66	67	69	72
− 5	2100-2249	30	51	62	64	65	68	70
− 10	2250-2399	26	49	60	62	64	66	69
− 15	2400-2549	23	46	58	60	62	64	67

Notes to Tables 22 and 23: *(1) Glass: double pane—0.45 Btu/ft^2·hr·°F; fibrous glass batt—0.3 Btu/ft^2·hr·°F/in.; urethane foam—0.133 Btu/ft^2·hr·°F/in. (2) Efficiency defined as ratio between energy actually transmitted through covers to energy transferred to storage, expressed as a percentage. (3) Low-temperature collector:Mean Absorber Temperature (T_i + T_o/2) = 90° F. (4) Average (daily, over 273-day heating season) energy transmitted through covers, assuming full-day random distribution of cloudiness: double cover, with no reflector—130 Btu/ft^2·hr; double cover with reflector at horizontal—155 Btu/ft^2·hr. (5) 15% side area assumed per ft^2*

Table 24 AVERAGE PERCENTAGE EFFICIENCY OF DOUBLE-COVER MID-TEMPERATURE COLLECTOR WITHOUT REFLECTORS (FOR DOMESTIC HOT WATER)

Outside Temp. (F)	Degree Days Per Month	Glass-Fiber Batt Insulation					Urethane Foam	
		R-1.67, 1/2"	R-3.33, 1"	R-6.67, 2"	R-8.33, 2-1/2"	R-10, 3"	R-15, 2"	R-30, 4"
65	0- 149	34	54	64	66	67	70	72
60	150- 299	30	51	62	64	65	68	70
55	300- 449	25	48	59	62	63	66	68
50	450- 599	21	45	57	59	61	64	66
45	600- 749	17	42	55	57	59	62	64
40	750- 899	12	39	52	55	57	59	62
35	900-1049	08	36	50	53	54	57	61
30	1050-1199	04	33	47	50	52	55	59
25	1200-1349	0	30	45	48	50	53	57
20	1350-1499		27	43	46	48	51	55
15	1500-1649		24	40	43	46	49	53
10	1650-1799		21	38	41	44	47	51
5	1800-1949		17	35	39	41	45	49
0	1950-2099		14	33	37	39	43	47
— 5	2100-2249		11	31	34	37	41	46
— 10	2250-2399		08	28	32	35	39	44
— 15	2400-2549		05	26	30	33	37	42

Table 25 AVERAGE PERCENTAGE EFFICIENCY OF DOUBLE-COVER MID-TEMPERATURE COLLECTOR WITH REFLECTORS (FOR DOMESTIC HOT WATER)

Outside Temp. (F)	Degree Days Per Month	Glass-Fiber Batt Insulation					Urethane Foam	
		R-1.67, 1/2"	R-3.33, 1"	R-6.67, 2"	R-8.33, 2-1/2"	R-10, 3"	R-15, 2"	R-30, 4"
65	0- 149	45	62	70	72	73	75	76
60	150- 299	41	59	68	70	71	73	75
55	300- 449	37	56	66	68	69	71	73
50	450- 599	34	54	64	66	67	69	72
45	600- 749	30	51	62	64	65	68	70
40	750- 899	26	49	60	62	64	66	68
35	900-1049	23	46	58	60	62	64	67
30	1050-1199	19	44	56	58	60	63	65
25	1200-1349	15	41	54	56	58	61	64
20	1350-1499	12	38	52	54	56	59	62
15	1500-1649	08	36	50	53	54	58	61
10	1650-1799	04	33	48	51	53	56	59
5	1800-1949	01	31	46	49	51	54	57
0	1950-2099	0	28	44	47	49	52	56
— 5	2100-2249		26	42	45	47	51	54
— 10	2250-2399		23	40	43	45	49	53
— 15	2400-2549		21	38	41	44	47	51

Notes to Tables 24 and 25: *(1) Glass: double pane—0.45 Btu/ft^2·hr·°F; fibrous glass batt—0.3 Btu/ft^2·hr· °F/ in.; urethane foam—0.133 Btu/ft^2·hr· °F/in. (2) Efficiency defined as ratio between energy actually transmitted through covers to energy transferred to storage, expressed as a percentage. (3) Mid-temperature collector = Mean Absorber Temperature ($T_i + T_o/2$) = 140 °F. (4) Average (daily, over 273-day heating season) energy transmitted through covers, assuming full-day random distribution of cloudiness: double cover, with no reflector—130 Btu/ft^2·hr; double cover with reflector at horizontal—155 Btu/ft^2·hr. (5) 15% side area assumed per ft^2.*

Table 26 AVERAGE PERCENTAGE EFFICIENCY OF DOUBLE-COVER HIGH-TEMPERATURE COLLECTOR WITHOUT REFLECTORS

Outside Temp. (F)	Degree Days Per Month	Glass-Fiber Batt Insulation					Urethane Foam	
		R-1.67, 1/2"	R-3.33, 1"	R-6.67, 2"	R-8.33, 2-1/2"	R-10, 3"	R-15, 2"	R-30, 4"
65	0- 149	04	33	47	50	52	55	59
60	150- 299	0	30	45	48	50	53	57
55	300- 449		27	43	46	48	51	55
50	450- 599		24	40	43	46	49	54
45	600- 749		21	38	41	44	47	52
40	750- 899		17	35	39	41	45	50
35	900-1049		14	33	37	39	43	48
30	1050-1199		11	31	34	37	42	46
25	1200-1349		08	28	32	35	39	44
20	1350-1499		05	26	30	33	37	42
15	1500-1649		02	23	28	30	35	41
10	1650-1799		0	21	25	28	33	39
5	1800-1949			19	23	26	31	37
0	1950-2099			16	21	24	29	35
5	2100-2249			14	19	22	27	33
10	2250-2399			11	16	20	25	31
15	2400-2549			09	14	17	23	29

Table 27 AVERAGE PERCENTAGE EFFICIENCY OF DOUBLE-COVER HIGH-TEMPERATURE COLLECTOR WITH REFLECTORS

Outside Temp. (F)	Degree Days Per Month	Glass-Fiber Batt Insulation					Urethane Foam	
		R-1.67, 1/2"	R-3.33, 1"	R-6.67, 2"	R-8.33, 2-1/2"	R-10, 3"	R-15, 2"	R-30, 4"
65	0- 149	19	44	56	58	60	63	66
60	150- 299	15	41	54	56	58	61	64
55	300- 449	12	38	52	54	56	59	63
50	450- 599	08	36	50	53	54	58	61
45	600- 749	04	33	48	51	53	56	59
40	750- 899	01	31	46	49	51	54	58
35	900-1049	0	28	44	47	49	52	56
30	1050-1199		26	42	45	47	51	55
25	1200-1349		23	40	43	45	49	53
20	1350-1499		21	38	41	44	47	52
15	1500-1649		18	36	39	42	46	50
10	1650-1799		15	34	37	40	44	49
5	1800-1949		13	32	36	38	42	47
0	1950-2099		10	30	34	36	41	45
5	2100-2249		08	28	32	34	39	44
10	2250-2399		05	26	30	33	37	42
15	2400-2549		03	24	28	31	35	41

Notes to Tables 26 and 27: (1) Glass: double pane—0.45 $Btu/ft^2 \cdot hr \cdot °F$; fibrous glass batt—0.3 $Btu/ft^2 \cdot hr \cdot °F/$ in.; urethane foam—0.133 $Btu/ft^2 \cdot hr \cdot °F/in$. (2) Efficiency defined as ratio between energy actually transmitted through covers to energy transferred to storage, expressed as a percentage. (3) High-temperature collector = Mean Absorber Temperature $(T_i + T_o/2) = 175 °F$. (4) Average (daily, over 273-day heating season) energy transmitted through covers, assuming full-day random distribution of cloudiness: double cover with no reflector—130 $Btu/ft^2 \cdot hr$; double cover with reflector at horizontal—155 $Btu/ft^2 \cdot hr$. (5) 15% side area assumed per ft^2

Table 28 AVERAGE PERCENTAGE EFFICIENCY OF ABOVE-GROUND LOW-TEMPERATURE STORAGE UNITS

Outside Temp. (F)	Degree Days Per Month	Glass-Fiber Batt Insulation				Urethane Foam	
		R-3.33, 1"	R-6.67, 2"	R-8.33, 2-1/2"	R-10, 3"	R-15, 2"	R-30, 4"
65	0- 149	33	66	73	78	85	93
60	150- 299	19	60	68	73	82	91
55	300- 449	06	53	62	69	79	90
50	450- 599	0	46	57	64	76	88
45	600- 749		39	51	60	73	87
40	750- 899		33	46	55	70	85
35	900-1049		26	41	51	67	84
30	1050-1199		19	35	46	64	82
25	1200-1349		12	30	42	61	81
20	1350-1499		06	24	37	58	79
15	1500-1649		0	19	33	55	78
10	1650-1799			14	28	52	76
5	1800-1949			08	24	49	75
0	1950-2099			03	19	46	73
5	2100-2249			0	15	43	72
10	2250-2399				10	40	70
15	2400-2549				06	37	69

Table 29 AVERAGE PERCENTAGE EFFICIENCY OF UNDERGROUND LOW-TEMPERATURE STORAGE UNITS

Outside Temp. (F)	Degree Days Per Month	Glass-Fiber Batt Insulation				Urethane Foam	
		R-3.33, 1"	R-6.67, 2"	R-8.33, 2-1/2"	R-10, 3"	R-15, 2"	R-30, 4"
All Months		06	53	62	69	79	90

Table 30 AVERAGE PERCENTAGE EFFICIENCY OF INDOOR LOW-TEMPERATURE STORAGE UNITS

Outside Temp. (F)	Degree Days Per Month	Glass-Fiber Batt Insulation				Urethane Foam	
		R-3.33, 1"	R-6.67, 2"	R-8.33, 2-1/2"	R-10, 3"	R-15, 2"	R-30, 4"
All Months		46	73	78	82	88	94

Notes to Tables 28-30: *(1) Assume 800 Btu/ft^2· day transferred from collector to storage. (2) Ratio of 1 ft^2 collector to 3 ft^2 wall area of storage. (3) Mean storage temperatue, 90° F. (4) Fibrous glass—0.3 Btu/ft^2 hr °F/in.; urethane foam—0.133 Btu/ft^2·hr·°F/in. (5) Ground temperature assumed 55°F. (6) Indoor temperature assumed 70°F. (7) 24-hour storage time.*

Table 31 AVERAGE PERCENTAGE EFFICIENCY OF ABOVE-GROUND MID-TEMPERATURE STORAGE UNITS

Outside Temp. (F)	Degree Days Per Month	Glass-Fiber Batt Insulation				Urethane Foam	
		R-3.33, 1"	R-6.67, 2"	R-8.33, 2-1/2"	R-10, 3"	R-15, 2"	R-30, 4"
65	0- 149	0	0	19	33	55	78
60	150- 299			14	28	52	76
55	300- 449			08	24	49	75
50	450- 599			03	19	46	73
45	600- 749			0	15	43	72
40	750- 899				10	40	70
35	900-1049				06	37	69
30	1050-1199				01	34	67
25	1200-1349				0	31	66
20	1350-1499					28	64
15	1500-1649					25	63
10	1650-1799					22	61
5	1800-1949					19	60
0	1950-2099					16	58
5	2100-2249					13	57
10	2250-2399					10	55
15	2400-2549					07	54

Table 32 AVERAGE PERCENTAGE EFFICIENCY OF UNDERGROUND MID-TEMPERATURE STORAGE UNITS

Outside Temp. (F)	Degree Days Per Month	Glass-Fiber Batt Insulation				Urethane Foam	
		R-3.33, 1"	R-6.67, 2"	R-8.33, 2-1/2"	R-10, 3"	R-15, 2"	R-30, 4"
All Months		0	0	08	24	49	75

Table 33 AVERAGE PERCENTAGE EFFICIENCY OF INDOOR MID-TEMPERATURE STORAGE UNITS

Outside Temp. (F)	Degree Days Per Month	Glass-Fiber Batt Insulation				Urethane Foam	
		R-3.33, 1"	R-6.67, 2"	R-8.33, 2-1/2"	R-10, 3"	R-15, 2"	R-30, 4"
All Months		0	06	24	37	58	79

Notes to Tables 31-33: *(1) Assume 800 Btu/ft^2·day transferred from collector to storage. (2) Ratio of 1 ft^2 collector to 3 ft^2 wall area of storage. (3) Mean storage temperature, 140°F. (4) Fibrous glass—0.3 Btu/ ft^2·hr·°F/in.; urethane foam—0.133 Btu/ft^2·hr·°F/in. (5) Ground temperature assumed 55°F. (6) Indoor temperature assumed 70°F. (7) 24-hour storage time.*

Table 34 AVERAGE PERCENTAGE EFFICIENCY OF ABOVE-GROUND HIGH-TEMPERATURE STORAGE UNITS

Outside Temp. (F)	Degree Days Per Month	Glass-Fiber Batt Insulation				Urethane Foam	
		R-3.33, 1"	R-6.67, 2"	R-8.33, 2-1/2"	R-10, 3"	R-15, 2"	R-30, 4"
65	0- 149	0	0	0	15	43	72
60	150- 299				10	40	70
55	300- 449				06	37	69
50	450- 599				01	34	67
45	600- 749				0	31	66
40	750- 899					28	64
35	900-1049					25	63
30	1050-1199					22	61
25	1200-1349					19	60
20	1350-1499					16	58
15	1500-1649					13	57
10	1650-1799					10	55
5	1800-1949					07	54
0	1950-2099					04	52
5	2100-2249					01	51
10	2250-2399					0	49
15	2400-2549						48

Table 35 AVERAGE PERCENTAGE EFFICIENCY OF UNDERGROUND HIGH-TEMPERATURE STORAGE UNITS

Outside Temp. (F)	Degree Days Per Month	Glass-Fiber Batt Insulation				Urethane Foam	
		R-3.33, 1"	R-6.67, 2"	R-8.33, 2-1/2"	R-10, 3"	R-15, 2"	R-30, 4"
All Months		0	0	0	06	37	69

Table 36 AVERAGE PERCENTAGE EFFICIENCY OF INDOOR HIGH-TEMPERATURE STORAGE UNITS

Outside Temp. (F)	Degree Days Per Month	Glass-Fiber Batt Insulation				Urethane Foam	
		R-3.33, 1"	R-6.67, 2"	R-8.33, 2-1/2"	R-10, 3"	R-15, 2"	R-30, 4"
All Months		0	0	03	19	46	73

Notes to Tables 34-36: *(1) Assume 800 Btu/ft^2 · day transferred from collector to storage. (2) Ratio of 1 ft^2 collector to 3 ft^2 wall area of storage. (3) Mean storage temperature, 160°F. (4) Fibrous glass—0.3 Btu/ft^2·hr·°F/in.; urethane foam—0.133 Btu/ft^2·hr·°F/in. (5) Ground temperature assumed 55°F. (6) Indoor temperature assumed 70°F. (7) 24-hour storage time.*

Table 37 B.T.U.'S TRANSMITTED TO INSIDE OF SINGLE-COVER COLLECTORS AT 60° TILT, WITHOUT REFLECTORS, PER SQUARE FOOT

Degrees North Latitude	Sep	Oct	Nov	Dec	Jan	Feb	Mar	Apr	May	Jun	Jul	Aug
25	1502	1793	1915	1931	1954	1969	1698	1298	975	576	785	1082
26	1519	1797	1903	1914	1938	1958	1712	1323	999	661	834	1122
27	1536	1801	1891	1897	1922	1947	1727	1348	1023	746	883	1162
28	1553	1806	1880	1879	1907	1936	1741	1372	1048	831	932	1201
29	1570	1810	1868	1862	1891	1925	1756	1397	1072	916	981	1241
30	1587	1814	1856	1845	1875	1914	1770	1422	1096	1001	1030	1281
31	1601	1813	1840	1823	1857	1907	1779	1442	1119	1024	1053	1304
32	1614	1812	1823	1800	1839	1900	1788	1462	1141	1046	1076	1327
33	1628	1811	1807	1778	1821	1894	1798	1482	1164	1069	1100	1351
34	1641	1810	1790	1755	1803	1887	1807	1502	1186	1091	1123	1374
35	1655	1809	1774	1733	1785	1880	1816	1522	1209	1114	1146	1397
36	1664	1802	1753	1700	1758	1867	1820	1545	1252	1136	1187	1416
37	1673	1795	1732	1666	1731	1853	1824	1569	1294	1158	1227	1435
38	1681	1788	1711	1633	1704	1840	1829	1592	1337	1181	1268	1453
39	1690	1781	1690	1599	1677	1826	1833	1616	1379	1203	1308	1472
40	1699	1774	1669	1566	1650	1813	1837	1639	1422	1225	1349	1491
41	1703	1761	1638	1520	1611	1794	1832	1653	1443	1267	1371	1507
42	1707	1748	1607	1474	1572	1775	1827	1667	1463	1309	1393	1522
43	1710	1734	1575	1427	1534	1756	1822	1681	1484	1351	1414	1538
44	1714	1721	1544	1381	1495	1737	1817	1695	1504	1393	1436	1553
45	1718	1708	1513	1335	1456	1718	1812	1709	1525	1435	1458	1569
46	1717	1690	1470	1284	1406	1693	1809	1718	1541	1455	1476	1588
47	1716	1673	1426	1232	1356	1668	1806	1728	1558	1475	1494	1607
48	1714	1655	1383	1181	1306	1642	1803	1737	1574	1494	1512	1627
49	1713	1638	1339	1129	1256	1617	1800	1747	1591	1514	1530	1646
50	1712	1620	1296	1078	1206	1592	1797	1756	1607	1534	1548	1665
51	1705	1590	1243	1010	1148	1556	1783	1761	1629	1551	1563	1673
52	1698	1561	1190	941	1090	1519	1769	1765	1651	1567	1577	1681
53	1692	1531	1138	873	1031	1483	1756	1770	1674	1584	1592	1690
54	1685	1502	1085	804	973	1446	1742	1774	1696	1600	1606	1698
55	1678	1472	1032	736	915	1410	1728	1779	1718	1617	1621	1706
56	1671	1442	979	668	857	1374	1714	1784	1740	1634	1636	1714
57	1664	1413	926	599	799	1337	1700	1788	1762	1650	1650	1722

Notes to Transmissivity Tables (37-44): *(1) Adapted from* Tables of Monthly Terrestrial Solar Radiation and Transmissivity, Single-and Double-Cover Collectors at 60° and 90° Tilts, With and Without Reflectors,

Table 38 B.T.U.'S TRANSMITTED TO INSIDE OF SINGLE-COVER COLLECTORS AT 60° TILT, WITH REFLECTORS, PER SQUARE FOOT

Degrees North Latitude	Sep	Oct	Nov	Dec	Jan	Feb	Mar	Apr	May	Jun	Jul	Aug
25	1537	1826	2084	2226	2189	1981	1725	1333	1019	625	835	1127
26	1556	1838	2093	2230	2196	2035	1739	1358	1043	711	884	1167
27	1573	1849	2102	2233	2203	2039	1754	1383	1067	797	933	1207
28	1587	1862	2111	2236	2211	2044	1767	1407	1093	882	983	1246
29	1604	1873	2120	2239	2218	2048	1782	1432	1117	968	1032	1286
30	1621	1885	2129	2243	2225	2052	1796	1457	1141	1054	1081	1326
31	1633	1899	2134	2234	2224	2065	1809	1477	1164	1077	1104	1349
32	1645	1912	2138	2224	2223	2078	1823	1497	1186	1098	1127	1372
33	1657	1926	2143	2214	2221	2092	1837	1517	1209	1121	1151	1397
34	1669	1939	2147	2204	2220	2105	1851	1537	1231	1142	1174	1420
35	1681	1953	2152	2195	2219	2118	1864	1557	1254	1165	1197	1443
36	1696	1965	2144	2154	2196	2127	1880	1580	1297	1187	1238	1462
37	1710	1977	2136	2111	2173	2135	1897	1604	1339	1209	1278	1481
38	1724	1989	2129	2070	2151	2143	1914	1627	1383	1232	1320	1499
39	1738	2001	2121	2027	2128	2151	1931	1651	1425	1254	1360	1518
40	1753	2013	2113	1986	2105	2160	1947	1674	1468	1276	1401	1537
41	1768	2020	2073	1920	2048	2158	1959	1689	1489	1318	1423	1553
42	1783	2027	2033	1855	1992	2157	1972	1705	1509	1360	1445	1568
43	1797	2033	1993	1788	1936	2155	1984	1720	1530	1402	1466	1584
44	1812	2040	1953	1723	1880	2154	1997	1736	1550	1444	1488	1599
45	1827	2047	1913	1657	1823	2152	2009	1751	1571	1486	1510	1615
46	1841	2045	1852	1585	1753	2124	2027	1768	1587	1506	1528	1636
47	1855	2044	1789	1513	1683	2097	2044	1785	1604	1526	1546	1657
48	1869	2042	1728	1441	1613	2068	2061	1802	1620	1545	1564	1678
49	1883	2041	1665	1369	1543	2041	2079	1819	1637	1565	1582	1699
50	1897	2039	1604	1297	1473	2013	2097	1836	1653	1585	1600	1720
51	1908	2001	1531	1208	1394	1960	2102	1854	1677	1602	1615	1736
52	1919	1964	1458	1119	1315	1906	2107	1870	1701	1618	1630	1752
53	1932	1927	1385	1030	1235	1853	2114	1888	1726	1635	1645	1768
54	1943	1890	1312	941	1156	1799	2119	1904	1750	1651	1660	1784
55	1954	1852	1239	852	1077	1746	2124	1922	1774	1668	1675	1800
56	1965	1814	1166	763	998	1693	2129	1940	1798	1685	1691	1816
57	1976	1777	1039	674	919	1639	2134	1956	1822	1701	1705	1832

Keyes, J.H., Morgan & Morgan, Dobbs Ferry, N.Y., 1979. (2) Transmissivities used may be seen in Appendix E. (3) Equations and definitions of calculations may be seen in Appendix H. (4) Horizontal reflectors

Table 39 B.T.U.'S TRANSMITTED TO INSIDE OF SINGLE-COVER COLLECTORS AT 90° TILT, WITHOUT REFLECTORS, PER SQUARE FOOT

Degrees North Latitude	Sep	Oct	Nov	Dec	Jan	Feb	Mar	Apr	May	Jun	Jul	Aug
25	587	1057	1390	1502	1476	1243	803	322	159	154	160	239
26	617	1081	1399	1505	1481	1264	835	353	175	160	170	261
27	647	1105	1407	1508	1487	1285	867	384	190	166	179	284
28	678	1128	1416	1511	1492	1306	900	414	206	171	189	306
29	708	1152	1424	1514	1498	1327	932	445	221	177	198	329
30	738	1176	1433	1517	1503	1348	964	476	237	183	208	351
31	767	1195	1437	1514	1504	1362	993	507	259	199	226	379
32	796	1214	1440	1511	1505	1376	1021	538	281	214	245	407
33	826	1234	1444	1508	1506	1391	1050	569	304	230	263	436
34	855	1253	1447	1505	1507	1405	1078	600	326	245	282	464
35	884	1272	1451	1502	1508	1419	1107	631	348	261	300	492
36	910	1285	1449	1486	1500	1426	1132	662	376	283	324	520
37	936	1298	1447	1471	1492	1433	1157	693	405	305	348	549
38	963	1312	1444	1455	1483	1440	1181	725	433	328	372	577
39	989	1325	1442	1440	1475	1447	1206	756	462	350	396	606
40	1015	1338	1440	1424	1467	1454	1231	787	490	372	420	634
41	1038	1344	1426	1392	1444	1455	1250	815	519	399	447	663
42	1061	350	1412	1360	1421	1455	1269	844	548	426	475	692
43	1083	1357	1398	1328	1399	1456	1287	872	576	454	502	722
44	1106	1363	1384	1296	1376	1456	1306	901	605	481	530	751
45	1129	1369	1370	1264	1353	1457	1325	929	634	508	557	780
46	1147	1370	1341	1224	1316	1450	1338	955	663	536	585	806
47	1165	1370	1312	1184	1279	1443	1350	981	693	564	613	832
48	1183	1371	1283	1145	1241	1436	1363	1008	722	592	641	859
49	1201	1371	1254	1105	1204	1429	1375	1034	752	602	669	885
50	1219	1372	1225	1065	1167	1422	1388	1060	781	648	697	911
51	1230	1361	1184	1003	1119	1401	1394	1083	809	677	725	934
52	1241	1349	1142	942	1070	1381	1400	1106	836	706	754	958
53	1253	1338	1101	880	1022	1360	1406	1128	864	735	782	981
54	1264	1326	1059	819	973	1340	1412	1151	891	764	811	1005
55	1275	1315	1018	757	925	1319	1418	1174	919	793	839	1028
56	1286	1304	977	695	877	1298	1424	1197	947	822	867	1051
57	1297	1292	935	634	828	1278	1430	1220	974	851	896	1075

used in all calculations with exception of summer months on 90°-tilt collectors. A 20° reflector tilt (ψ_r) is used for 90°-tilt collectors at all latitudes during May, June and July; and for North Latitudes 25-48 during

Table 40 B.T.U.'S TRANSMITTED TO INSIDE OF SINGLE-COVER COLLECTORS AT 90° TILT, WITH REFLECTORS, PER SQUARE FOOT

Degrees North Latitude	Sep	Oct	Nov	Dec	Jan	Feb	Mar	Apr	May	Jun	Jul	Aug
25	1,357	1,672	2,233	2,327	2,334	1,990	1,244	1,054	574	403	469	820
26	1,381	1,716	2,231	2,310	2,321	2,017	1,300	1,094	624	448	517	867
27	1,406	1,760	2,228	2,292	2,308	2,064	1,356	1,133	674	492	563	915
28	1,431	1,804	2,226	2,275	2,295	2,102	1,412	1,172	724	536	611	962
29	1,456	1,848	2,223	2,257	2,282	2,139	1,468	1,211	774	580	657	1,010
30	1,480	1,892	2,221	2,240	2,269	2,176	1,524	1,251	824	625	705	1,057
31	1,499	1,926	2,206	2,214	2,249	2,185	1,575	1,281	871	673	752	1,092
32	1,518	1,960	2,190	2,188	2,228	2,194	1,625	1,311	918	719	800	1,128
33	1,537	1,995	2,174	2,162	2,208	2,204	1,676	1,341	965	767	848	1,164
34	1,556	2,029	2,158	2,136	2,187	2,213	1,726	1,371	1,012	813	896	1,200
35	1,575	2,063	2,143	2,110	2,167	2,222	1,777	1,401	1,059	861	943	1,235
36	1,588	2,067	2,119	2,069	2,135	2,211	1,822	1,425	1,097	906	984	1,262
37	1,601	2,071	2,096	2,029	2,103	2,201	1,867	1,449	1,135	950	1,025	1,290
38	1,614	2,075	2,071	1,989	2,071	2,190	1,912	1,475	1,173	996	1,065	1,317
39	1,627	2,079	2,048	1,949	2,039	2,180	1,957	1,499	1,211	1,040	1,106	1,345
40	1,640	2,083	2,024	1,908	2,007	2,169	2,002	1,523	1,249	1,085	1,147	1,372
41	1,680	2,072	1,986	1,850	1,959	2,150	2,022	1,540	1,278	1,121	1,178	1,395
42	1,720	2,061	1,948	1,793	1,911	2,129	2,042	1,559	1,307	1,156	1,211	1,417
43	1,758	2,051	1,911	1,735	1,863	2,110	2,061	1,576	1,334	1,193	1,242	1,441
44	1,798	2,040	1,873	1,678	1,815	2,089	2,081	1,595	1,363	1,228	1,275	1,463
45	1,838	2,029	1,835	1,620	1,767	2,070	2,101	1,612	1,392	1,264	1,306	1,486
46	1,860	2,011	1,781	1,556	1,705	2,040	2,101	1,624	1,416	1,292	1,331	1,502
47	1,881	1,991	1,728	1,491	1,643	2,011	2,099	1,636	1,441	1,320	1,357	1,518
48	1,903	1,973	1,674	1,428	1,580	1,981	2,099	1,649	1,464	1,348	1,382	1,534
49	1,924	1,953	1,621	1,363	1,518	1,952	2,097	1,677	1,489	1,358	1,408	1,550
50	1,946	1,935	1,567	1,299	1,456	1,922	2,097	1,723	1,513	1,404	1,433	1,566
51	1,943	1,902	1,502	1,215	1,385	1,877	2,085	1,763	1,532	1,428	1,454	1,576
52	1,941	1,868	1,436	1,131	1,312	1,833	2,073	1,803	1,549	1,452	1,476	1,586
53	1,939	1,834	1,372	1,047	1,241	1,787	2,060	1,842	1,568	1,476	1,498	1,596
54	1,937	1,800	1,306	963	1,168	1,743	2,048	1,882	1,585	1,500	1,520	1,631
55	1,934	1,767	1,241	879	1,097	1,698	2,036	1,922	1,604	1,524	1,541	1,672
56	1,931	1,734	1,176	795	1,026	1,653	2,024	1,962	1,623	1,548	1,562	1,713
57	1,929	1,700	1,110	711	953	1,609	2,012	2,002	1,640	1,572	1,584	1,755

April, North Latitudes 25-53 during August and North Latitudes 25-40 during September. (5) *All values given for 15th day of month and per net square foot of collector cover area.*

Table 41 B.T.U.'S TRANSMITTED TO INSIDE OF DOUBLE-COVER COLLECTORS AT 60° TILT, WITHOUT REFLECTORS, PER SQUARE FOOT

Degrees North Latitude	Sep	Oct	Nov	Dec	Jan	Feb	Mar	Apr	May	Jun	Jul	Aug
25	1273	1541	1664	1684	1702	1658	1449	1095	808	474	648	909
26	1288	1546	1654	1669	1689	1658	1463	1117	831	545	690	942
27	1304	1551	1645	1655	1676	1659	1476	1140	853	616	732	976
28	1319	1555	1635	1640	1662	1659	1490	1162	876	688	774	1009
29	1335	1560	1626	1626	1649	1660	1503	1185	898	759	816	1043
30	1350	1565	1616	1611	1636	1660	1517	1207	921	830	858	1076
31	1363	1564	1602	1592	1620	1655	1526	1225	942	851	879	1097
32	1375	1563	1588	1572	1605	1650	1535	1243	963	872	901	1119
33	1388	1562	1575	1553	1589	1644	1544	1261	983	893	922	1140
34	1400	1561	1561	1533	1574	1639	1553	1279	1004	914	944	1162
35	1413	1560	1547	1514	1558	1634	1562	1297	1025	935	965	1183
36	1422	1556	1529	1485	1535	1623	1567	1316	1061	955	999	1200
37	1431	1552	1511	1456	1511	1612	1572	1336	1097	976	1033	1217
38	1439	1547	1492	1427	1488	1601	1576	1355	1133	996	1067	1233
39	1448	1543	1474	1398	1464	1590	1581	1375	1169	1017	1101	1250
40	1457	1539	1456	1369	1441	1579	1586	1394	1205	1037	1135	1267
41	1461	1528	1429	1329	1407	1563	1586	1407	1223	1073	1155	1281
42	1465	1517	1402	1289	1373	1547	1586	1420	1241	1108	1175	1295
43	1470	1506	1375	1248	1340	1530	1585	1434	1258	1144	1194	1310
44	1474	1495	1348	1208	1306	1514	1585	1447	1276	1179	1214	1324
45	1478	1484	1321	1168	1272	1498	1585	1460	1294	1215	1234	1338
46	1478	1469	1283	1123	1229	1476	1579	1469	1309	1233	1250	1354
47	1478	1454	1246	1078	1185	1454	1574	1478	1325	1250	1267	1370
48	1477	1438	1208	1034	1142	1432	1568	1488	1340	1268	1283	1386
49	1477	1423	1171	989	1098	1410	1563	1497	1356	1285	1300	1402
50	1477	1408	1133	944	1055	1388	1557	1506	1371	1303	1316	1418
51	1471	1383	1087	884	1004	1356	1545	1511	1389	1318	1330	1426
52	1466	1358	1041	824	954	1325	1533	1516	1408	1333	1343	1434
53	1460	1332	995	764	903	1293	1522	1520	1426	1349	1357	1443
54	1455	1307	949	704	853	1262	1510	1525	1445	1364	1370	1451
55	1449	1282	903	644	802	1230	1498	1530	1463	1379	1384	1459
56	1443	1257	857	584	751	1198	1486	1535	1481	1394	1398	1467
57	1438	1232	811	524	701	1167	1474	1540	1500	1409	1411	1475

Table 42 B.T.U.'S TRANSMITTED TO INSIDE OF DOUBLE-COVER COLLECTORS
AT 60° TILT, WITH REFLECTORS, PER SQUARE FOOT

Degrees North Latitude	Sep	Oct	Nov	Dec	Jan	Feb	Mar	Apr	May	Jun	Jul	Aug
25	1300	1566	1782	1908	1874	1699	1470	1122	842	512	687	944
26	1315	1575	1790	1912	1881	1710	1484	1144	865	583	729	977
27	1331	1585	1798	1917	1889	1721	1497	1167	887	655	771	1011
28	1346	1593	1806	1920	1895	1732	1511	1189	911	727	813	1044
29	1362	1603	1814	1925	1903	1743	1524	1212	933	799	855	1078
30	1377	1612	1822	1929	1910	1754	1538	1234	956	871	897	1111
31	1390	1621	1827	1923	1910	1765	1549	1252	977	892	918	1132
32	1402	1631	1832	1916	1911	1777	1561	1270	998	913	940	1154
33	1415	1640	1838	1910	1911	1787	1572	1288	1018	933	962	1175
34	1427	1650	1843	1903	1912	1799	1584	1306	1039	953	984	1197
35	1439	1659	1848	1897	1912	1810	1595	1324	1060	974	1005	1218
36	1451	1671	1843	1863	1895	1811	1608	1343	1096	994	1039	1235
37	1462	1683	1838	1828	1877	1812	1621	1363	1132	1015	1073	1252
38	1473	1694	1833	1794	1859	1814	1634	1382	1169	1036	1107	1269
39	1484	1706	1828	1759	1841	1815	1647	1402	1205	1057	1141	1286
40	1495	1718	1823	1725	1824	1816	1660	1421	1241	1077	1175	1303
41	1506	1725	1790	1669	1776	1824	1674	1435	1259	1113	1195	1317
42	1518	1732	1758	1613	1728	1831	1688	1449	1277	1148	1215	1331
43	1530	1738	1725	1557	1682	1838	1700	1463	1294	1184	1234	1346
44	1542	1745	1693	1501	1634	1845	1714	1477	1312	1219	1254	1360
45	1553	1752	1660	1445	1586	1853	1728	1491	1330	1255	1274	1374
46	1565	1752	1607	1382	1526	1831	1740	1505	1345	1273	1290	1391
47	1578	1752	1555	1320	1465	1809	1753	1518	1361	1290	1307	1408
48	1589	1752	1502	1258	1406	1787	1765	1533	1376	1308	1323	1425
49	1602	1753	1450	1196	1345	1765	1778	1546	1392	1325	1340	1442
50	1614	1753	1397	1133	1285	1743	1790	1560	1407	1343	1356	1459
51	1624	1723	1334	1055	1216	1697	1796	1575	1426	1358	1370	1472
52	1634	1693	1271	977	1148	1653	1801	1589	1446	1373	1383	1485
53	1644	1663	1208	897	1079	1607	1808	1603	1465	1389	1398	1498
54	1654	1633	1145	822	1011	1563	1813	1617	1485	1404	1411	1511
55	1664	1603	1082	744	942	1517	1819	1632	1504	1419	1425	1524
56	1674	1573	1019	666	873	1471	1825	1647	1523	1434	1439	1537
57	1684	1543	956	588	805	1427	1830	1661	1543	1449	1453	1550

Table 43 B.T.U.'S TRANSMITTED TO INSIDE OF DOUBLE-COVER COLLECTORS AT 90° TILT, WITHOUT REFLECTORS, PER SQUARE FOOT

Degrees North

Latitude	Sep	Oct	Nov	Dec	Jan	Feb	Mar	Apr	May	Jun	Jul	Aug
25	689	1390	1918	2014	2013	1681	993	343	174	174	180	249
26	735	1432	1918	2000	2003	1716	1045	384	189	179	189	276
27	780	1473	1918	1985	1994	1752	1096	426	205	185	198	303
28	827	1514	1917	1972	1983	1788	1149	466	219	189	205	330
29	872	1555	1917	1957	1974	1824	1200	508	235	195	214	357
30	918	1597	1917	1943	1964	1859	1252	549	250	200	223	384
31	966	1630	1905	1921	1947	1869	1299	594	277	220	242	422
32	1012	1662	1892	1899	1930	1880	1345	638	304	239	263	461
33	1060	1696	1880	1876	1913	1890	1393	684	330	260	282	499
34	1106	1728	1867	1854	1896	1901	1439	728	357	279	303	538
35	1154	1761	1855	1832	1879	1911	1486	733	384	299	322	576
36	1196	1766	1835	1798	1851	1903	1529	821	423	323	353	618
37	1238	1771	1815	1763	1825	1895	1571	869	461	346	385	659
38	1281	1776	1795	1728	1797	1888	1613	917	501	371	417	701
39	1323	1781	1775	1693	1771	1880	1655	965	539	394	449	742
40	1365	1786	1755	1659	1743	1872	1698	1013	578	418	480	784
41	1403	1779	1722	1610	1702	1857	1717	1058	620	456	519	828
42	1441	1770	1691	1561	1662	1840	1737	1103	662	493	558	873
43	1479	1763	1658	1513	1621	1825	1756	1147	704	530	597	918
44	1517	1754	1627	1464	1581	1808	1776	1192	746	567	636	963
45	1555	1747	1594	1415	1540	1793	1795	1237	788	605	675	1007
46	1576	1731	1549	1360	1486	1767	1795	1279	834	647	717	1048
47	1596	1717	1504	1305	1434	1743	1796	1321	878	689	759	1090
48	1616	1701	1457	1248	1380	1717	1796	1364	924	729	802	1132
49	1636	1687	1412	1193	1328	1693	1797	1406	968	771	844	1174
50	1657	1671	1367	1138	1274	1667	1797	1448	1014	813	886	1215
51	1656	1643	1310	1064	1211	1629	1788	1484	1057	857	929	1253
52	1655	1615	1255	992	1149	1591	1778	1521	1100	901	972	1290
53	1654	1588	1198	918	1087	1555	1769	1557	1143	946	1015	1329
54	1653	1560	1143	846	1025	1517	1759	1594	1186	990	1058	1366
55	1652	1532	1086	772	962	1479	1750	1630	1229	1034	1101	1404
56	1651	1504	1029	698	899	1441	1741	1666	1272	1078	1144	1442
57	1650	1476	974	626	837	1403	1731	1703	1315	1122	1187	1479

Table 44 B.T.U.'S TRANSMITTED TO INSIDE OF DOUBLE-COVER COLLECTORS AT 90° TILT, WITH REFLECTORS, PER SQUARE FOOT

Degrees North Latitude	Sep	Oct	Nov	Dec	Jan	Feb	Mar	Apr	May	Jun	Jul	Aug
25	1115	1390	1918	2014	2013	1681	993	862	475	330	387	669
26	1137	1432	1918	2000	2003	1716	1045	894	515	367	426	708
27	1160	1473	1918	1985	1994	1752	1096	928	555	404	464	746
28	1183	1514	1917	1972	1983	1788	1149	960	595	441	502	786
29	1206	1555	1917	1957	1974	1824	1200	994	635	478	540	824
30	1228	1597	1917	1943	1964	1859	1252	1026	675	515	579	863
31	1247	1630	1905	1921	1947	1869	1299	1053	713	553	617	893
32	1266	1662	1892	1899	1930	1880	1345	1079	752	590	655	924
33	1285	1696	1880	1876	1913	1890	1393	1107	789	629	693	953
34	1304	1728	1867	1854	1896	1901	1439	1133	828	666	731	984
35	1323	1761	1855	1832	1879	1911	1486	1160	866	704	769	1014
36	1336	1766	1835	1798	1851	1903	1529	1182	898	741	803	1038
37	1348	1771	1815	1763	1825	1895	1571	1206	930	778	837	1063
38	1362	1776	1795	1728	1797	1888	1613	1228	962	815	871	1087
39	1374	1781	1775	1693	1771	1880	1655	1252	994	852	905	1112
40	1387	1786	1755	1659	1743	1872	1698	1274	1026	889	939	1136
41	1403	1779	1722	1610	1702	1857	1717	1291	1051	919	967	1157
42	1441	1770	1691	1561	1662	1840	1737	1308	1077	949	994	1178
43	1479	1763	1658	1513	1621	1825	1756	1324	1103	979	1022	1199
44	1517	1754	1627	1464	1581	1808	1776	1341	1129	1009	1049	1220
45	1555	1747	1594	1415	1540	1793	1795	1358	1154	1039	1077	1241
46	1576	1731	1549	1360	1486	1767	1795	1370	1177	1064	1100	1256
47	1596	1717	1504	1305	1434	1743	1796	1382	1198	1090	1123	1271
48	1616	1701	1457	1248	1380	1717	1796	1394	1221	1114	1147	1288
49	1636	1687	1412	1193	1328	1693	1797	1406	1242	1140	1170	1303
50	1657	1671	1367	1138	1274	1667	1797	1448	1265	1165	1193	1318
51	1656	1643	1310	1064	1211	1629	1788	1484	1282	1187	1212	1328
52	1655	1615	1255	992	1149	1591	1778	1521	1299	1209	1231	1338
53	1654	1588	1198	918	1087	1555	1769	1557	1317	1230	1251	1348
54	1653	1560	1143	846	1025	1517	1759	1594	1334	1252	1270	1366
55	1652	1532	1086	772	962	1479	1750	1630	1351	1274	1289	1404
56	1651	1504	1029	698	899	1441	1741	1666	1368	1296	1308	1442
57	1650	1476	974	626	837	1403	1731	1703	1385	1318	1327	1479

Table 45 POSSIBLE HOURS OF SUNSHINE PER MONTH AT VARIOUS LATITUDES

Degrees North Latitude	Sep	Oct	Nov	Dec	Jan	Feb	Mar	Apr	May
25	369	359	328	329	333	317	371	382	414
26	369	359	327	327	331	316	371	383	416
27	370	358	325	324	329	315	371	384	417
28	370	357	323	322	327	314	371	385	420
29	371	356	321	319	325	313	371	386	421
30	371	355	320	317	323	312	371	387	423
31	371	355	318	315	321	311	371	388	425
32	372	354	316	312	319	310	370	389	427
33	372	353	314	310	316	309	370	390	429
34	372	352	313	307	314	308	370	391	431
35	372	351	311	305	312	307	370	392	433
36	373	350	309	302	309	306	370	393	436
37	373	349	306	299	307	304	370	394	439
38	373	349	304	295	304	303	369	396	441
39	374	348	302	293	302	301	369	397	443
40	374	347	300	290	299	300	369	398	446
41	375	346	297	286	296	299	369	399	449
42	375	345	295	283	293	297	369	400	452
43	376	344	293	280	290	295	369	402	455
44	377	343	290	275	287	294	368	403	457
45	377	342	288	273	284	293	368	405	460
46	378	340	285	269	280	291	368	406	464
47	378	339	282	264	276	289	368	408	467
48	379	338	278	260	272	287	368	410	471
49	379	336	275	255	269	285	367	411	474
50	380	335	272	251	265	283	367	413	477
51	380	333	268	246	260	280	367	415	482
52	381	332	264	240	255	278	366	417	486
53	381	330	260	235	251	276	366	419	491
54	382	329	256	229	246	274	365	421	495
55	383	327	252	224	241	272	365	423	500
56	384	325	248	219	236	270	365	425	505

Notes: *(1) Based on possible hours of sunshine on the 15th day of the month. (2) Derived from tables,* "Total Possible Sunshine," Climatic Atlas of the United States, *Environmental Data Service, U.S. Department of Commerce, June 1968.*

Table 46 MEAN PERCENTAGE OF POSSIBLE SUNSHINE

	Sep	Oct	Nov	Dec	Jan	Feb	Mar	Apr	May
ALABAMA									
Birmingham	66	67	58	44	43	49	56	63	66
Huntsville	65	68	58	42	43	48	55	62	65
Mobile	63	72	61	46	49	51	57	65	71
Montgomery	69	71	64	48	51	53	61	69	73
ARIZONA									
Flagstaff	80	78	70	70	60	70	73	76	80
Phoenix	89	88	84	77	76	79	83	88	93
Prescott	85	82	82	75	71	75	79	82	87
Tucson	90	90	89	83	81	86	86	90	93
Winslow	84	83	81	73	68	75	80	83	86
Yuma	93	93	90	83	83	87	91	94	97
ARKANSAS									
Fort Smith	70	66	56	48	47	51	55	60	62
Little Rock	71	74	58	47	44	53	57	62	67
Texarkana	73	76	62	53	50	55	59	63	69
CALIFORNIA									
Bakersfield	90	85	78	55	55	70	75	80	85
Bishop	86	70	58	55	65	65	70	70	80
Burbank	83	74	75	70	70	72	75	70	65
Eureka	52	48	42	39	40	44	50	53	54
Fresno	93	87	73	47	46	63	72	83	89
Long Beach	83	74	75	70	70	72	75	70	65
Los Angeles	80	76	79	72	70	69	70	67	68
Oakland	80	72	60	50	50	58	70	72	78
Red Bluff	89	77	64	50	50	60	65	75	79
Sacramento	92	82	65	44	44	57	67	76	82
San Diego	70	70	76	71	68	67	68	66	60
San Francisco	70	70	62	54	53	57	63	69	70
Santa Maria	78	73	71	65	64	72	70	68	65
COLORADO									
Alamosa	75	60	62	54	60	60	58	55	60
Colorado Springs	72	70	65	70	70	70	70	62	61
Denver	71	71	67	65	67	67	65	63	61
Grand Junction	77	74	67	58	58	62	64	67	71
Pueblo	78	76	74	71	74	72	71	68	68
CONNECTICUT									
Bridgeport	62	55	52	50	50	51	52	54	62
Hartford	57	55	46	46	46	55	56	54	57
New Haven	63	62	53	54	52	60	58	59	61

	Sep	*Oct*	*Nov*	*Dec*	*Jan*	*Feb*	*Mar*	*Apr*	*May*
DELAWARE									
Wilmington	62	61	53	49	45	56	57	58	61
DISTRICT OF COLUMBIA									
Washington	62	61	54	47	46	53	56	57	61
FLORIDA									
Appalachicola	62	74	66	53	59	62	62	71	77
Daytona Beach	55	58	62	60	62	64	66	71	69
Jacksonville	58	58	61	53	58	59	66	71	71
Key West	65	65	69	66	68	75	78	78	76
Miami Beach	62	62	65	65	66	72	73	73	68
Orlando	58	60	62	62	63	66	69	72	72
Pensacola	67	75	65	53	55	58	63	70	73
Tallahassee	61	70	65	56	56	60	65	70	73
Tampa	64	67	67	61	63	67	71	74	75
GEORGIA									
Athens	69	70	62	50	51	54	61	66	70
Atlanta	65	67	60	47	48	53	57	65	68
Augusta	70	72	61	52	52	54	62	70	71
Columbus	63	72	62	50	52	55	62	67	73
Macon	68	67	64	54	56	58	64	72	75
Rome	63	65	58	44	45	50	55	62	67
Savannah	57	61	62	54	55	56	62	70	72
IDAHO									
Boise	81	66	46	37	40	48	59	68	68
Idaho Falls	75	60	42	31	34	42	52	60	61
Lewiston	71	60	33	19	29	39	50	62	62
Pocatello	78	66	48	36	37	47	58	64	66
ILLINOIS									
Cairo	75	73	56	46	46	53	59	65	71
Chicago	65	61	47	41	44	49	53	56	63
Moline	67	62	44	43	45	47	51	54	56
Peoria	69	64	51	43	46	50	54	57	60
Rockford	66	61	45	42	44	45	54	55	61
Springfield	73	64	53	45	47	51	54	58	64
INDIANA									
Evansville	73	67	52	42	42	49	55	61	67
Fort Wayne	64	58	41	38	38	44	51	55	62
Indianapolis	68	64	48	39	41	47	49	55	62
South Bend	64	59	38	35	37	41	52	54	63
IOWA									
Burlington	72	70	59	51	50	55	59	60	63
Des Moines	64	64	53	48	56	56	56	59	62

	Sep	Oct	Nov	Dec	Jan	Feb	Mar	Apr	May
Dubuque	61	55	44	40	48	52	52	58	60
Sioux City	67	64	53	48	56	59	58	62	63
Waterloo	66	62	48	45	52	55	57	58	62
KANSAS									
Concordia	72	70	64	58	60	60	62	63	65
Dodge City	76	75	70	67	67	66	68	68	68
Goodland	73	71	66	66	68	69	68	66	67
Topeka	70	66	57	51	53	53	52	54	59
Wichita	73	69	67	59	61	63	64	64	66
KENTUCKY									
Covington	68	60	46	39	41	46	52	56	62
Lexington	65	59	48	36	38	43	49	55	60
Louisville	68	64	51	39	41	47	52	57	64
LOUISIANA									
Alexandria	72	75	63	50	47	52	56	67	68
Baton Rouge	69	72	61	48	49	50	57	61	67
Lake Charles	70	75	62	48	48	52	55	62	69
New Orleans	64	70	60	46	49	50	57	63	66
Shreveport	79	77	65	60	48	54	58	60	69
MAINE									
Caribou	42	42	28	28	32	40	41	44	45
Portland	61	59	50	54	54	59	58	56	59
MARYLAND									
Baltimore	64	61	54	49	49	56	57	58	61
Frederick	62	60	50	45	42	51	52	52	61
MASSACHUSETTS									
Boston	61	58	48	48	47	56	57	56	59
Nantucket	65	60	50	45	43	52	58	57	62
Pittsfield	50	50	38	31	41	50	51	51	54
Worcester	58	54	44	44	43	53	54	53	57
MICHIGAN									
Alpena	52	44	24	22	29	43	52	56	59
Detroit	61	54	35	29	34	42	48	52	58
Escanaba	52	48	32	35	40	51	55	56	57
Flint	58	50	30	27	32	49	51	53	58
Grand Rapids	58	50	31	29	28	40	48	54	60
Lansing	59	50	32	28	32	41	51	54	61
Marquette	47	38	24	24	31	40	47	52	53
Muskegon	59	48	30	28	29	39	51	55	59
Sault Ste. Marie	45	36	21	22	28	44	50	54	54
MINNESOTA									
Duluth	53	47	36	40	47	55	60	58	58

	Sep	Oct	Nov	Dec	Jan	Feb	Mar	Apr	May
International Falls	54	48	34	41	45	52	54	57	58
Minneapolis	60	54	40	40	49	54	55	57	60
Rochester	63	61	42	41	50	54	54	57	61
Saint Cloud	59	55	38	42	49	54	53	58	59
MISSISSIPPI									
Jackson	63	63	59	48	41	47	54	63	66
Meridian	68	67	59	47	49	50	58	65	69
Vicksburg	74	71	60	45	46	50	57	64	69
MISSOURI									
Columbia	70	65	55	47	49	54	56	58	63
Kansas City	70	67	59	52	55	57	59	60	64
St. Joseph	70	65	56	50	52	55	57	58	61
St. Louis	67	65	54	44	48	49	56	59	64
Springfield	71	65	58	48	48	54	57	60	63
MONTANA									
Billings	68	63	48	48	50	53	57	58	61
Glasgow	65	60	48	46	50	58	60	62	63
Great Falls	68	61	47	51	57	61	67	64	63
Havre	64	57	48	46	49	58	61	63	63
Helena	63	57	48	43	46	55	58	60	59
Kalispell	61	50	28	20	28	40	49	57	58
Miles City	66	63	52	51	52	61	61	63	64
Missoula	65	53	32	25	31	38	45	51	56
NEBRASKA									
Grand Island	71	67	59	55	60	62	62	62	65
Lincoln	67	66	59	55	57	59	60	60	63
Norfolk	68	64	55	50	57	60	60	61	64
North Platte	72	70	62	58	63	63	64	62	64
Omaha	72	72	56	51	58	63	60	65	68
Scottsbluff	70	68	60	61	63	66	65	62	62
NEVADA									
Elko	83	72	60	52	56	60	62	63	67
Ely	81	73	67	62	61	64	68	65	67
Las Vegas	92	84	83	75	74	77	78	81	85
Reno	86	76	68	56	59	64	69	75	77
Winnemucca	86	75	62	53	52	60	64	70	76
NEW HAMPSHIRE									
Concord	55	50	43	43	48	53	55	53	51
NEW JERSEY									
Atlantic City	65	54	58	52	51	57	58	59	62
Newark	64	61	53	50	49	56	57	59	62
Trenton	64	60	53	49	48	56	55	59	62

	Sep	Oct	Nov	Dec	Jan	Feb	Mar	Apr	May
NEW MEXICO									
Albuquerque	81	80	79	70	70	72	72	76	79
Raton	72	70	70	62	70	72	70	65	65
Roswell	74	74	74	69	69	72	75	77	76
Silver City	85	81	81	75	72	80	82	82	81
NEW YORK									
Albany	58	54	39	38	43	51	53	53	57
Binghampton	47	43	29	26	31	39	41	44	50
Buffalo	60	51	31	28	32	41	49	51	59
New York City	64	61	53	50	49	56	57	59	62
Rochester	60	50	33	31	32	42	47	52	60
Schenectady	58	54	39	38	43	51	53	53	57
Syracuse	56	47	29	26	31	38	45	50	58
NORTH CAROLINA									
Asheville	62	64	59	48	48	53	56	61	64
Charlotte	66	69	64	55	53	58	62	68	72
Greensboro	65	67	61	54	51	56	59	59	68
Raleigh	63	64	62	52	50	56	59	64	67
Winston-Salem	63	53	58	51	48	52	56	62	65
NORTH DAKOTA									
Bismarck	62	59	49	48	52	58	56	57	58
Devils Lake	59	56	44	45	53	60	59	60	59
Fargo	60	57	39	46	47	55	56	58	62
Williston	65	60	48	48	51	59	60	63	66
OHIO									
Akron	62	55	33	25	29	36	45	52	62
Cincinnati	68	60	46	39	41	46	52	56	62
Cleveland	62	54	32	25	29	36	45	52	61
Columbus	66	60	44	35	36	44	49	54	63
Dayton	72	66	51	43	38	45	53	56	63
Mansfield	64	56	38	30	33	41	48	52	62
Sandusky	66	58	38	32	34	43	50	57	63
Toledo	62	55	38	30	32	43	49	53	60
Youngstown	61	52	32	25	28	39	45	50	60
OKLAHOMA									
Oklahoma City	74	68	64	57	57	60	63	64	65
Tulsa	75	69	67	60	49	54	54	54	56
OREGON									
Astoria	45	41	27	22	27	32	40	48	48
Eugene	64	42	28	20	24	33	40	50	50
Medford	66	50	38	25	28	39	48	52	55
Pendleton	74	60	38	21	35	38	50	61	71

	Sep	Oct	Nov	Dec	Jan	Feb	Mar	Apr	May
Portland	55	42	28	23	27	34	41	49	52
Roseburg	68	42	28	18	24	32	40	51	57
Salem	60	42	28	20	26	33	40	50	50
PENNSYLVANIA									
Allentown	62	59	49	46	44	54	56	57	61
Erie	61	49	29	25	28	39	46	49	58
Harrisburg	62	58	47	43	43	52	55	57	61
Philadelphia	62	61	53	49	45	56	57	58	61
Pittsburgh	62	54	39	30	32	39	45	50	57
Reading	59	57	48	44	44	50	53	55	58
Scranton	57	53	40	37	36	46	48	50	56
Williamsport	55	50	35	31	34	42	48	49	54
RHODE ISLAND									
Providence	62	60	50	44	48	58	54	55	60
SOUTH CAROLINA									
Charleston	67	68	68	57	58	60	65	72	73
Columbia	64	68	64	51	53	57	62	68	69
Florence	63	67	63	50	52	56	61	67	68
Greenville	64	66	62	51	53	57	61	70	71
Spartanburg	62	64	59	48	48	53	56	61	64
SOUTH DAKOTA									
Huron	66	61	52	49	55	62	60	62	65
Rapid City	69	66	58	54	58	62	63	62	61
Sioux Falls	67	65	53	50	55	58	58	59	63
TENNESSEE									
Bristol	59	59	49	38	39	42	49	58	59
Chattanooga	66	63	54	42	40	48	51	61	67
Knoxville	64	64	53	41	42	49	53	59	64
Memphis	70	69	58	45	44	51	57	64	68
Nashville	69	65	55	42	42	47	54	60	65
TEXAS									
Abilene	73	71	72	66	64	68	73	66	73
Amarillo	79	76	76	70	71	71	75	75	75
Austin	70	70	57	49	46	50	57	60	62
Brownsville	67	70	54	44	44	49	51	57	65
Corpus Christi	75	74	60	51	49	53	57	62	70
Dallas	74	70	63	58	56	57	65	66	67
El Paso	80	82	80	73	74	77	81	85	87
Fort Worth	74	70	63	58	56	57	65	66	67
Galveston	70	74	62	49	50	50	55	61	69
Houston	64	67	57	46	45	45	52	55	63
Laredo	74	70	64	55	55	60	65	65	65

	Sep	Oct	Nov	Dec	Jan	Feb	Mar	Apr	May
Lubbock	77	77	77	70	72	72	75	75	77
Midland	80	78	78	70	73	70	75	75	79
Port Arthur	68	72	60	47	47	48	56	61	69
San Angelo	73	71	71	66	64	67	72	66	73
San Antonio	69	67	55	49	48	51	56	58	60
Victoria	69	74	60	49	50	50	55	63	69
Waco	72	70	60	53	51	53	61	63	65
Wichita Falls	74	69	64	57	57	58	64	65	66
UTAH									
Salt Lake City	84	73	56	49	48	53	61	68	73
VERMONT									
Burlington	51	43	25	24	34	43	48	47	53
VIRGINIA									
Lynchburg	63	62	58	53	50	56	59	61	65
Norfolk	63	64	60	51	50	57	60	63	67
Richmond	63	64	58	50	49	55	59	63	67
Roanoke	60	60	49	41	40	41	50	60	60
WASHINGTON									
Olympia	54	37	29	25	27	34	43	49	54
Seattle	53	36	28	24	27	34	42	48	53
Spokane	68	53	28	22	26	41	53	63	64
Tacoma	53	36	28	24	27	34	42	48	53
Walla Walla	72	59	33	20	24	35	51	63	67
Yakima	74	61	38	29	34	49	62	70	72
WEST VIRGINIA									
Charleston	58	54	38	32	33	38	44	48	54
Elkins	55	51	41	33	33	37	42	47	55
Huntington	60	56	40	33	34	40	45	50	58
Parkersburg	60	53	37	29	30	36	42	49	56
WISCONSIN									
Green Bay	58	52	40	40	44	51	55	56	58
La Crosse	62	58	43	39	49	51	53	58	59
Madison	60	56	41	38	44	49	52	53	58
Milwaukee	62	56	44	39	44	48	53	56	60
WYOMING									
Casper	70	63	56	56	62	66	68	65	61
Cheyenne	69	69	65	63	65	66	64	61	59
Lander	72	67	61	62	66	70	71	66	65
Sheridan	67	60	53	52	56	61	62	61	61
ALBERTA									
Banff	45	39	31	17	23	33	34	39	41
Beaverlodge	46	43	33	28	31	39	44	51	54

	Sep	*Oct*	*Nov*	*Dec*	*Jan*	*Feb*	*Mar*	*Apr*	*May*
Brooks	52	52	41	33	34	42	42	50	56
Calgary	49	50	43	38	38	43	43	47	49
Edmonton	49	49	40	34	36	41	48	53	55
Edson	43	45	30	29	34	42	42	45	48
Fairview	43	40	29	23	27	36	45	51	52
Lacombe	48	47	38	33	34	41	45	48	51
Lethbridge	57	53	42	38	40	46	47	50	55
Medicine Hat	49	49	39	34	34	42	41	48	54
Olds	49	50	40	34	35	43	43	45	48
Vauxhall	54	51	44	37	37	43	44	48	56
BRITISH COLUMBIA									
Campbell River	43	29	20	11	18	31	34	38	53
Dawson Creek	45	42	33	27	31	39	44	50	55
Fort St. John	45	42	33	27	31	39	44	50	55
Hope	39	27	15	13	16	24	26	28	35
Kamloops	50	34	21	13	19	30	41	40	47
Kitimat	32	19	14	13	19	24	34	37	43
Nanaimo	50	35	22	18	20	32	35	40	50
Nelson	50	31	20	10	13	29	38	37	51
New Westminster	48	35	25	18	18	32	36	40	52
Penticton	54	38	23	15	17	30	41	47	53
Port Alberni	50	35	22	18	20	32	35	40	50
Powell River	45	33	22	19	15	32	35	39	52
Prince George	41	32	23	17	22	32	38	44	52
Prince Rupert	25	16	16	10	17	22	24	28	32
Quesnel	41	32	23	17	22	32	38	44	52
Summerland	54	38	22	15	17	29	41	46	52
Terrace	32	19	14	13	19	24	34	37	43
Trail	50	31	14	5	9	25	35	34	45
Vancouver	48	35	25	18	18	32	36	40	52
Victoria	55	41	29	24	26	34	41	48	58
MANITOBA									
Brandon	49	48	31	34	40	45	45	48	50
Dauphin	46	47	33	37	45	49	48	52	55
Morden	50	50	32	36	40	48	44	49	50
Rivers	51	51	35	38	44	50	47	51	55
The Pas	44	43	20	28	32	43	45	47	51
Winnipeg	48	47	30	34	42	49	60	51	52
NEW BRUNSWICK									
Campbellton	44	37	28	32	35	43	42	44	48
Chatham	48	42	32	36	40	42	40	44	45
Fredericton	44	41	30	34	37	41	38	39	43

	Sep	Oct	Nov	Dec	Jan	Feb	Mar	Apr	May
Moncton	44	41	31	33	37	41	37	41	46
Saint John	46	45	31	40	39	43	42	39	42
NEWFOUNDLAND									
Gander	38	33	23	24	27	30	28	28	33
St. John's	40	32	23	21	25	27	25	28	36
NOVA SCOTIA									
Annapolis Royal	47	41	26	19	22	29	38	39	45
Halifax	47	45	32	31	33	40	39	40	44
Kentville	46	41	26	21	26	33	35	39	45
Sydney	44	41	26	25	29	36	34	40	44
Truro	40	37	26	25	30	36	34	40	44
Yarmouth	45	43	29	21	25	32	37	43	50
ONTARIO									
Atikokan	49	31	23	27	38	54	54	47	45
Belleville	46	44	27	29	29	37	37	40	45
Brampton	51	47	32	30	34	39	39	44	53
Chatham	51	48	30	25	27	33	34	41	48
Cornwall	46	40	26	24	32	40	43	45	50
Guelph	45	42	26	25	28	34	36	42	50
Hamilton	45	41	25	26	32	41	38	43	54
Harrow	51	47	28	25	26	33	34	41	51
Kapuskasing	32	27	15	20	28	36	39	40	40
Kingston	50	45	31	29	34	40	39	43	55
Lindsay	46	41	25	21	27	36	39	42	46
London	47	44	25	22	24	33	35	42	51
Mount Forest	46	41	20	22	25	40	39	46	54
New Liskeard	35	29	17	19	26	36	42	41	37
North Bay	42	34	21	26	35	45	43	46	50
Ottawa	45	40	26	29	34	39	41	43	50
St. Catharines	49	40	26	18	20	26	34	40	47
Sarnia	49	49	24	21	27	36	36	47	52
Sault Ste. Marie	42	36	22	25	27	39	43	46	53
Thunder Bay	46	35	29	36	43	52	50	50	50
Toronto	52	45	28	28	30	37	39	44	48
Woodstock	51	46	26	23	24	31	36	43	48
PRINCE EDWARD ISLAND									
Charlottetown	48	39	25	22	30	36	37	38	43
QUEBEC									
Amos	36	27	16	27	29	40	42	45	46
Berthierville	49	44	27	32	35	41	46	48	50
Grandes Bergeronnes	44	38	26	28	35	36	43	46	43
L'Assomption	47	42	28	29	35	41	42	45	50

	Sep	*Oct*	*Nov*	*Dec*	*Jan*	*Feb*	*Mar*	*Apr*	*May*
Lennoxville	44	39	23	22	28	35	37	40	46
Maniwaki	42	37	23	26	34	43	42	49	52
Montreal	52	44	26	29	34	42	46	46	52
Normandin	38	32	23	29	36	40	43	44	45
Quebec	44	37	22	25	29	34	38	40	42
Rimouski	42	30	21	17	23	28	34	40	39
Ste. Agathe-des-Monts	45	40	21	30	34	43	41	49	53
St. Ambroise	37	27	20	28	37	40	42	49	45
Sept-Iles	44	37	32	38	38	47	44	54	49
Sherbrooke	46	41	25	24	29	35	38	42	47
Victoriaville	48	40	24	26	32	38	43	45	48
SASKATCHEWAN									
Estevan	59	57	42	42	46	48	53	52	59
Indian Head	55	49	36	33	40	46	48	50	55
Melfort	43	44	29	29	39	42	43	47	49
Moose Jaw	54	52	37	33	38	41	44	52	56
Outlook	47	47	35	32	39	42	46	49	57
Prince Albert	45	43	31	31	37	42	46	51	53
Regina	51	50	35	33	37	41	43	51	57
Saskatoon	54	53	37	35	39	46	52	54	57
Scott	48	45	32	26	33	41	41	50	56
Swift Current	51	50	39	34	36	41	41	50	55
Weyburn	58	46	42	35	41	43	49	52	60
Yorkton	49	35	28	36	39	45	47	51	56

Notes to Table of Mean Percentage of Possible Sunshine: *(1) Values for some U.S. cities from tables, "Mean Percentage of Possible Sunshine," and some from maps, "Mean Percentage of Possible Sunshine, September through May,"* Climatic Atlas of the United States, *Environmental Data Service, U.S. Department of Commerce, June 1968. (2) Values for some U.S. cities calculated by dividing values, "Mean Total Hours of Sunshine," same source as No. 1, by appropriate values for latitude of subject city from Table 45 (Possible Hours of Sunshine at Various Latitudes). (3) Values for Canadian cities obtained by procedure of No. 2 using "Daily Bright Sunshine," Yorke, B.J. and Kendall, G.R., Atmospheric Environment Service, Department of the Environment of Canada, Downsview, Ontario, 1972. (4) Values for some cities in British Columbia obtained by extrapolation and isoline mapping.*

SECTION SIX

How Long Will It Take to Pay for Itself in Fuel Savings on Your Home?

14 Determining the Pay-Back Time

PAY-BACK IS NOT A VALID CONCEPT
The solar furnace that is added to your present house, or the one that is built into your new house, becomes a fixed improvement to your real property. Therefore, it *appreciates* in value, right along with the rest of your house. It doesn't *depreciate;* so the concept of paying for itself in fuel savings is not a valid one. Unfortunately, as a result of long efforts by ignorant governmental bureaucrats and some not so ignorant, the idea has grabbed hold and is impossible to stop now.

As an analogy, suppose that you bought a tract house with an unfinished basement. Later you added a bathroom to the basement at a cost of about $1,000. If you were to call in an appraiser, the result would be an appraisal of your home at $1,000 more than your neighbor's, which does not have the additional bathroom. You certainly would hoot at anyone's suggesting that you should pay off in "flushes per month."

In the future it may be impossible to sell a house *without* solar heating equipment; but that day is still a long way off. In the meantime, with inflation and wages being the way they are, you are going to have to be *powerfully* motivated before you rush out to buy a solar furnace for your home. Especially when the going rate is somewhere between $2,500 and $5,000!

Even worse, you are going to have to be armed to the teeth with justifications when your brother-in-law comes over for the first time after you've installed it.

And just who are you to believe? The oil companies have got it made. We can't check their numbers to find out just how real the "energy crisis" is. There certainly is no doubt that we will run out of fossil fuels *sometime.* The question is, when? And, despite all their fine support of the arts and public and educational TV, the energy companies—who (to hear them tell it) put the good of humanity first—keep raising the prices of fuel until our senior citizens now face fuel and utility bills higher than their mortgage payments used to be. Leaving them with the choice of buying fuel *or* food. Heat or eat, but not both on that monthly pension check in midwinter.

No matter whom you believe about the energy crisis and how real it is, we can all believe this: Fuel is going to go up in price year by year, by 10% or 20% or however much the profiteers think they can gouge out of us while intoning solemn "tsk-tsks" to the media about how awful the inflation is and how much it costs to explore for new supplies.

THE SOLAR PAY-BACK-GRAM

Here, for the first time anywhere, is a simple estimator designed for you, to allow you to determine fairly accurately the length of time the solar heating equipment you purchase will take to pay for itself in fuel savings.

You will need to know the price per square foot of collector for the system you are buying. For example, suppose that you are considering the purchase of a system selling for $4,350 installed on your house. It has 128 square feet of collector area. That works out to be $33.98 per square foot. Because you aren't rich, you are going to have to finance it. Interest over the life of the loan is going to cost you an additional $910. Adding that in brings the cost per square foot of collector up to $41.09. If, on the other hand, you qualify for a tax credit from your state and/or the federal govern-

ment, you would subtract that from the total cost. A tax credit of $1,020 would bring the cost per square foot of collector down to $33.13.

At the end of this chapter is Table 30, "Heating Season Btu Output From One Square Foot of Collector." You will need to know the value for your home town or the nearest town to your home.

Finally, you will need to know how much you are paying for your fuel now. If you use natural gas, you will want to know the cost per thousand cubic feet; if propane or fuel oil, the cost per gallon, and if you heat with electricity, you will need the cost per kilowatt hour. When looking up that cost on your last heating bill, be sure to add in any fuel surcharge that your local utility has put on your bill, sometimes in an obscure place, as though it were not a rise in cost to you.

There are separate Collector-Solar-Pay-Back-Grams for each of the fuels; so be sure that you find the appropriate one for you.

Even though the wholesale price of new natural gas has risen 300% in the last year and other fuels have gone up by gross amounts, the cost of fuel has been assumed to escalate only at about the same rate as annual inflation—say, 10%. If fuel costs go up faster, as they probably will, your solar heating equipment will be paid off faster than shown on the Pay-Back-Gram.

Nomogram 3 SOLAR-PAY-BACK-GRAM FOR NATURAL GAS HEAT
(For determining the number of years needed to pay, in fuel savings, for solar heating equipment)

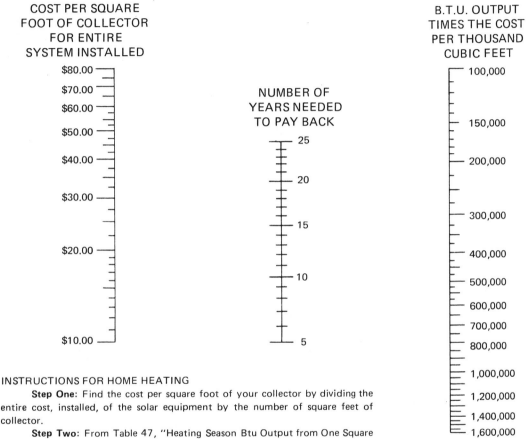

COST PER SQUARE
FOOT OF COLLECTOR
FOR ENTIRE
SYSTEM INSTALLED

NUMBER OF
YEARS NEEDED
TO PAY BACK

B.T.U. OUTPUT
TIMES THE COST
PER THOUSAND
CUBIC FEET

INSTRUCTIONS FOR HOME HEATING

Step One: Find the cost per square foot of your collector by dividing the entire cost, installed, of the solar equipment by the number of square feet of collector.

Step Two: From Table 47, "Heating Season Btu Output from One Square Foot of Collector," below, find the number of Btu's delivered inside your house, in your community.

Step Three: From last month's heating bill, find the cost that you paid, including fuel adjustment surcharges, per thousand cubic feet (MCF) of natural gas. (Some utilities use a hundred cubic feet (CCF) as the basis for their billing; in this case, multiply that rate by 10 to obtain the thousand-cubic-foot rate.) Multiply the cost per MCF times the number obtained in Step Two. Example: You pay $1.67 per MCF, and your town's Btu number is 223,000. 223,000 X 1.67 = 372,410.

Step Four: If your solar collector is to be ground-mounted, multiply answer obtained in Step Three by 1.2.

Step Five: Using a straightedge, line up the numbers obtained in Steps One and Three. The straightedge will intersect the center line at the approximate number of years required to pay for the equipment in fuel savings.

DOMESTIC HOT WATER HEATING Follow all steps for "Home Heating," above. However, after completing Step Three, multiply the total obtained by 1.2.

Notes to Solar-Pay-Back-Grams: *(1) References generally assume an efficiency of 70% for fuel-burning furnaces. This is barely possible in unrestricted high heat exchange (high flow rate) systems under laboratory*

Nomogram 4 SOLAR-PAY-BACK-GRAM FOR FUEL OIL
(For determining number of years needed to pay for solar equipment in savings on fuel oil)

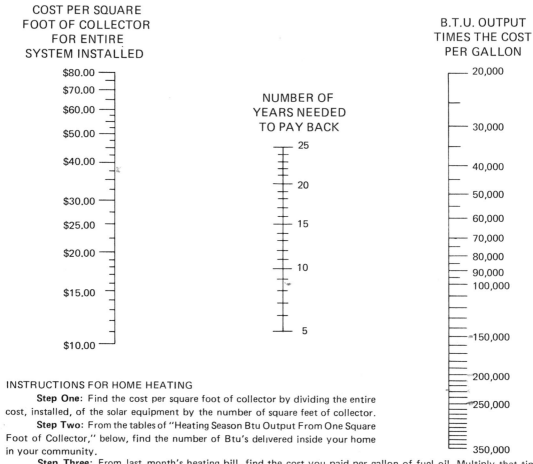

INSTRUCTIONS FOR HOME HEATING

Step One: Find the cost per square foot of collector by dividing the entire cost, installed, of the solar equipment by the number of square feet of collector.

Step Two: From the tables of "Heating Season Btu Output From One Square Foot of Collector," below, find the number of Btu's delivered inside your home in your community.

Step Three: From last month's heating bill, find the cost you paid per gallon of fuel oil. Multiply that times the number obtained in Step Two. Example: You paid 40 cents per gallon. There are 200,000 Btu's in your home town. 200,000 x .40 = 80,000

Step Four: If your solar collector is to be ground-mounted, multiply answer obtained in Step Three by 1.2.

Step Five: Using a straightedge, line up the numbers obtained in Steps One and Three. The straightedge will intersect the center line at the approximate number of years it will take to pay for the equipment in fuel savings.

DOMESTIC HOT WATER HEATING Follow all steps for "Home Heating," above. However, after completing Step Three, multiply the total obtained by 1.2.

conditions. Testing of a first-line name-brand furnace showed that 50% efficiency is optimistic in a typically restricted system installed in the real world, and that after typical use in the average home, 35-to-40% is probably a more realistic estimate. However, 50% efficiency has been used for these Solar-Pay-Back-Grams. (2) Values used: 1,000 cubic feet of natural gas = net 500,000 Btu's; 1 gallon fuel oil = net 72,000 Btu's; 1 gallon propane = net 46,000 Btu's; and 1 kilowatt hour = 3413 Btu's; all useful heat delivery in-

Nomogram 5 SOLAR-PAY-BACK-GRAM FOR PROPANE
(For determining the number of years needed to pay, in fuel savings, for solar equipment)

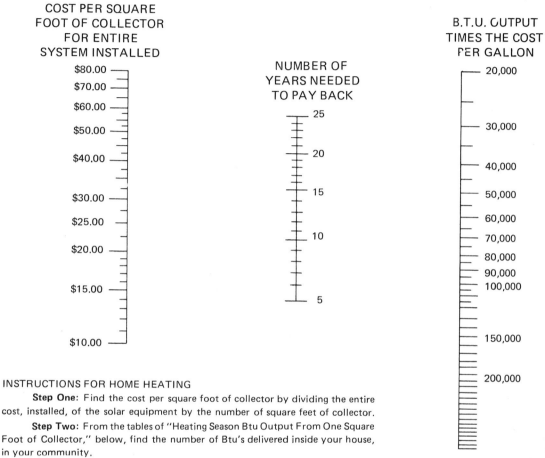

COST PER SQUARE
FOOT OF COLLECTOR
FOR ENTIRE
SYSTEM INSTALLED

NUMBER OF
YEARS NEEDED
TO PAY BACK

B.T.U. OUTPUT
TIMES THE COST
PER GALLON

INSTRUCTIONS FOR HOME HEATING

Step One: Find the cost per square foot of collector by dividing the entire cost, installed, of the solar equipment by the number of square feet of collector.

Step Two: From the tables of "Heating Season Btu Output From One Square Foot of Collector," below, find the number of Btu's delivered inside your house, in your community.

Step Three: From last month's heating bill, find the cost per gallon of propane. Multiply that by the number obtained in Step Two. Example: You paid 39.9 cents per gallon. There are 235,567 Btu's in your home town. 235,567 x .399 = 93,991.23. Rounding off gives you 94,000.

Step Four: If your solar collector is to be ground-mounted, multiply answer obtained in Step Three by 1.2.

Step Five: Using a straightedge, line up the numbers obtained in Steps One and Three. The straightedge will intersect the center line at the approximate number of years required to pay for the equipment in fuel savings.

DOMESTIC HOT WATER HEATING Follow all steps for "Home Heating," above. However, after completing Step Three, multiply the total obtained by 1.2.

side the home. (3) 90% of the 273-day heating season's possible output from storage assumed as useful; 10% assumed lost in higher heat production by solar equipment than home requires during September and May. (4) 10% per annum fuel cost escalation is assumed, with the exception of natural gas, which has been projected at 12% per annum cost escalation despite the fact that cost escalation has been higher, historically, for the past few years. (5) No real-estate appreciation (typically 5% per annum) has been imputed to the solar equipment for these Solar-Pay-Back-Grams, although it can be argued convincingly that solar equipment becomes real property improvement.

Nomogram 6 SOLAR-PAY-BACK-GRAM FOR ELECTRICAL HEAT
(For determining the number of years needed for electrical savings to pay for solar equipment)

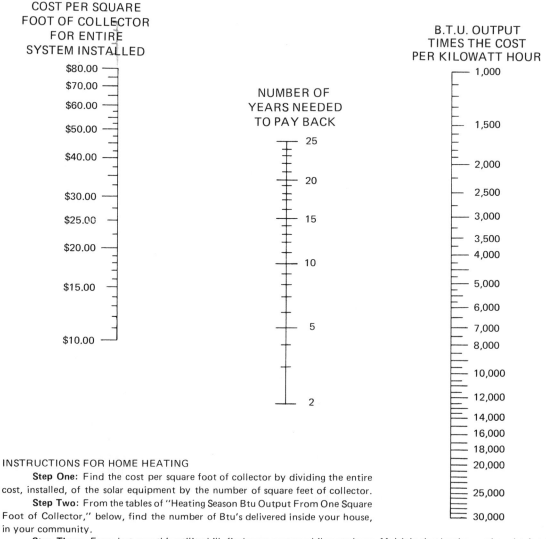

INSTRUCTIONS FOR HOME HEATING

 Step One: Find the cost per square foot of collector by dividing the entire cost, installed, of the solar equipment by the number of square feet of collector.

 Step Two: From the tables of "Heating Season Btu Output From One Square Foot of Collector," below, find the number of Btu's delivered inside your house, in your community.

 Step Three: From last month's utility bill, find your cost per kilowatt hour. Multiply that by the number obtained in Step Two. Example: You paid 3.5 cents per kwh, and the Btu number for your town is 300,000. 300,300 x .035 = 10,500.

 Step Four: If your solar collector is to be ground-mounted, multiply answer obtained in Step Three by 1.2.

 Step Five: Using a straightedge, line up the numbers obtained in Steps One and Three. Where the straightedge intersects the center line you will find the approximate number of years it will take to pay for the equipment in electrical savings.

DOMESTIC HOT WATER HEATING Follow all steps for "Home Heating," above. However, after completing Step Three, multiply the total obtained by 1.2.

196,000*

Table 47 HEATING SEASON B.T.U. OUTPUT FROM 1 SQUARE FOOT OF R-30-INSULATED DOUBLE-COVER COLLECTOR WITH HORIZONTAL REFLECTORS, R-30-INSULATED STORAGE

	Btu Output for 60°-Tilt Collector	Btu Output for 90°-Tilt Collector		Btu Output for 60°-Tilt Collector	Btu Output for 90°-Tilt Collector
ALABAMA			**CONNECTICUT**		
Birmingham	204,707	196,069	Bridgeport	186,856	179,266
Huntsville	201,262	193,744	Hartford	178,553	171,968
Mobile	213,726	202,252	New Haven	200,132	192,513
Montgomery	221,561	210,295	**DELAWARE**		
ARIZONA			Wilmington	197,578	190,017
Flagstaff	241,593	232,645	**DISTRICT OF COLUMBIA**		
Phoenix	309,253	296,925	Washington	196,456	188,904
Prescott	279,481	269,330	**FLORIDA**		
Tucson	321,065	307,645	Appalachicola	234,148	221,234
Winslow	273,557	263,332	Daytona Beach	230,637	190,299
Yuma	334,071	319,910	Jacksonville	223,845	211,958
ARKANSAS			Key West	251,995	229,478
Fort Smith	204,197	196,718	Miami Beach	241,853	223,330
Little Rock	211,821	204,160	Orlando	237,244	221,174
Texarkana	224,314	216,074	Pensacola	232,819	220,742
CALIFORNIA			Tallahassee	230,627	217,795
Bakersfield	273,480	263,490	Tampa	243,763	227,018
Bishop	242,339	232,767	**GEORGIA**		
Burbank	272,009	262,212	Athens	219,507	210,418
Eureka	168,920	162,273	Atlanta	210,106	201,401
Fresno	263,686	253,332	Augusta	225,420	215,103
Long Beach	271,461	261,736	Columbus	221,301	212,291
Los Angeles	270,279	260,800	Macon	232,256	221,555
Oakland	239,614	230,271	Rome	199,081	190,658
Red Bluff	249,318	239,377	Savannah	221,530	210,543
Sacramento	247,734	237,864	**IDAHO**		
San Diego	257,764	248,290	Boise	197,760	190,850
San Francisco	232,046	223,392	Idaho Falls	165,505	159,618
Santa Maria	257,302	249,339	Lewiston	164,496	158,902
COLORADO			Pocatello	156,532	150,632
Alamosa	197,325	189,834	**ILLINOIS**		
Colorado Springs	232,616	224,483	Cairo	214,012	205,659
Denver	229,853	221,583	Chicago	181,954	174,931
Grand Junction	229,468	220,590	Moline	177,234	170,580
Pueblo	251,056	242,167	Peoria	189,046	181,566

	Btu Output for 60°-Tilt Collector	Btu Output for 90°-Tilt Collector		Btu Output for 60°-Tilt Collector	Btu Output for 90°-Tilt Collector
Rockford	178,317	171,423	Pittsfield	153,730	148,218
Springfield	197,660	189,645	Worcester	170,462	164,098
INDIANA			**MICHIGAN**		
Evansville	197,916	176,095	Alpena	138,237	133,347
Fort Wayne	171,583	164,452	Detroit	156,354	150,091
Indianapolis	181,956	174,572	Escanaba	152,192	147,031
South Bend	167,163	160,465	Flint	150,017	144,392
IOWA			Grand Rapids	149,513	143,640
Burlington	206,405	198,451	Lansing	153,463	147,502
Des Moines	194,071	186,777	Marquette	127,173	122,712
Dubuque	173,566	166,691	Muskegon	149,188	143,427
Sioux City	200,140	190,999	Sault Ste. Marie	125,236	120,741
Waterloo	188,061	181,333	**MINNESOTA**		
KANSAS			Duluth	155,124	149,930
Concordia	223,370	214,829	International Falls	149,650	144,356
Dodge City	238,275	214,977	Minneapolis	169,025	163,378
Goodland	234,063	225,514	Rochester	174,333	168,310
Topeka	200,995	193,474	Saint Cloud	164,914	159,287
Wichita	229,173	205,463	**MISSISSIPPI**		
KENTUCKY			Jackson	199,664	189,432
Covington	183,131	175,656	Meridian	210,886	200,102
Lexington	175,587	156,225	Vicksburg	213,403	202,499
Louisville	187,799	167,340	**MISSOURI**		
LOUISIANA			Columbia	201,811	193,945
Alexandria	219,261	207,661	Kansas City	212,929	204,827
Baton Rouge	214,615	203,414	St. Joseph	203,775	195,899
Lake Charles	215,740	203,615	St. Louis	197,573	189,786
New Orleans	209,503	197,563	Springfield	204,024	196,247
Shreveport	229,868	220.090	**MONTANA**		
MAINE			Billings	186,615	180,659
Caribou	119,023	114,996	Glasgow	178,316	172,386
Portland	185,954	179,805	Great Falls	192,578	186,430
MARYLAND			Havre	176,978	170,955
Baltimore	199,646	190,669	Helena	175,673	170,014
Frederick	184,735	176,127	Kalispell	141,993	136,979
MASSACHUSETTS			Miles City	191,817	185,490
Boston	182,853	175,863	Missoula	143,772	138,982
Nantucket	189,344	181,993	**NEBRASKA**		

	Btu Output for 60°-Tilt Collector	Btu Output for 90°-Tilt Collector		Btu Output for 60°-Tilt Collector	Btu Output for 90°-Tilt Collector
Grand Island	211,698	203,504	NORTH DAKOTA		
Lincoln	209,286	201,262	Bismark	174,876	169,251
Norfolk	201,822	194,260	Devils Lake	170,075	164,430
North Platte	219,597	211,343	Fargo	168,270	162,646
Omaha	212,555	204,250	Williston	180,332	174,223
Scottsbluff	217,267	209,495	OHIO		
NEVADA			Akron	152,070	145,361
Elko	214,748	206,413	Cincinnati	182,365	175,024
Ely	225,232	216,919	Cleveland	150,414	143,750
Las Vegas	294,637	283,981	Columbus	173,668	166,402
Reno	240,184	230,656	Dayton	188,147	180,525
Winnemucca	225,324	216,165	Mansfield	160,846	153,956
NEW HAMPSHIRE			Sandusky	168,595	161,784
Concord	166,091	160,328	Toledo	159,204	152,796
NEW JERSEY			Youngstown	148,502	142,010
Atlantic City	203,438	195,335	OKLAHOMA		
Newark	200,590	192,764	Oklahoma City	224,927	216,829
Trenton	198,474	190,658	Tulsa	213,663	206,408
NEW MEXICO			OREGON		
Albuquerque	281,421	255,655	Astoria	126,886	122,502
Raton	234,138	226,097	Eugene	138,079	133,091
Roswell	256,384	247,151	Medford	156,193	149,959
Silver City	280,271	268,173	Pendleton	173,615	167,522
NEW YORK			Portland	137,271	132,581
Albany	167,563	161,531	Roseburg	141,162	135,380
Binghampton	132,069	126,759	Salem	136,663	132,013
Buffalo	150,272	144,629	PENNSYLVANIA		
New York City	200,538	192,675	Allentown	186,663	179,618
Rochester	153,384	147,417	Erie	145,156	139,141
Schenectady	167,563	161,531	Harrisburg	185,890	178,470
Syracuse	143,207	137,549	Philadelphia	196,709	189,180
NORTH CAROLINA			Pittsburgh	155,845	149,279
Asheville	201,681	194,254	Reading	180,547	173,240
Charlotte	225,804	217,552	Scranton	160,943	154,476
Greensboro	212,634	204,830	Williamsport	151,744	145,578
Raleigh	212,319	204,329	RHODE ISLAND		
Wilmington	233,629	224,056	Providence	187,833	180,944
Winston-Salem	199,759	191,719	SOUTH CAROLINA		

	Btu Output for 60°-Tilt Collector	Btu Output for 90°-Tilt Collector		Btu Output for 60°-Tilt Collector	Btu Output for 90°-Tilt Collector
Charleston	237,268	226,759	Burlington	134,219	129,707
Columbia	223,420	214,341	**VIRGINIA**		
Florence	219,766	210,824	Lynchburg	207,218	199,370
Greenville	222,027	213,656	Norfolk	213,945	205,938
Spartanburg	205,744	198,317	Richmond	209,082	200,993
SOUTH DAKOTA			Roanoke	180,535	173,234
Huron	192,554	186,184	**WASHINGTON**		
Rapid City	203,246	196,579	Olympia	133,710	129,094
Sioux Falls	192,530	185,918	Seattle	134,104	129,441
TENNESSEE			Spokane	156,978	151,380
Bristol	177,741	170,645	Tacoma	130,935	126,350
Chattanooga	193,601	186,059	Walla Walla	166,149	160,404
Knoxville	192,482	185,078	Yakima	186,586	180,135
Memphis	207,740	199,896	**WEST VIRGINIA**		
Nashville	195,908	188,217	Charleston	155,910	149,598
TEXAS			Elkins	152,654	146,258
Abilene	250,748	240,151	Huntington	162,816	156,121
Amarillo	263,194	253,808	Parkersburg	152,269	145,787
Austin	206,975	195,111	**WISCONSIN**		
Brownsville	195,462	178,091	Green Bay	164,763	159,340
Corpus Christi	218,207	202,483	La Crosse	173,463	167,369
Dallas	230,687	220,545	Madison	164,364	158,370
El Paso	284,904	271,316	Milwaukee	169,281	163,017
Fort Worth	230,687	220,545	**WYOMING**		
Galveston	216,340	201,491	Casper	209,060	201,984
Houston	197,376	186,233	Cheyenne	216,180	208,351
Laredo	229,472	213,764	Lander	219,854	212,438
Lubbock	263,962	253,625	Sheridan	193,402	187,283
Midland	269,837	257,318			
Port Arthur	210,213	197,992	**ALBERTA**		
San Angelo	249,287	236,919	Banff	102,754	99,526
San Antonio	205,525	194,105	Beaverlodge	118,235	114,693
Victoria	215,018	200,004	Brooks	136,305	131,849
Waco	219,878	209,186	Calgary	134,539	130,439
Wichita Falls	229,650	220,471	Edmonton	133,589	129,512
UTAH			Edson	116,431	112,880
Salt Lake City	215,355	206,473	Fairview	110,185	106,882
VERMONT			Lacombe	127,215	123,309

	Btu Output for 60°-Tilt Collector	Btu Output for 90°-Tilt Collector		Btu Output for 60°-Tilt Collector	Btu Output for 90°-Tilt Collector
Lethbridge	149,177	144,418	NEWFOUNDLAND		
Medicine Hat	135,079	130,604	Gander	91,364	88,516
Olds	131,074	124,499	St. John's	92,117	89,144
Vauxhall	142,807	138,217	NOVA SCOTIA		
BRITISH COLUMBIA			Annapolis Royal	113,486	109,795
Campbell River	104,105	100,406	Halifax	130,153	126,076
Dawson Creek	116,145	112,606	Kentville	114,183	110,366
Fort St. John	114,895	111,483	Sydney	115,532	111,648
Hope	84,480	81,591	Truro	113,214	109,304
Kamloops	109,762	106,237	Yarmouth	120,505	116,076
Kitimat	83,186	80,446	ONTARIO		
Nanaimo	114,051	110,015	Atikokan	122,184	118,164
Nelson	105,281	101,753	Belleville	122,635	118,346
New Westminster	115,175	111,011	Brampton	134,893	129,959
Penticton	119,756	115,398	Chatham	127,751	122,586
Port Alberni	113,182	109,157	Cornwall	125,060	120,739
Powell River	110,851	107,020	Guelph	119,880	115,367
Prince George	100,861	97,537	Hamilton	129,464	124,301
Prince Rupert	59,910	58,145	Harrow	127,133	126,765
Quesnel	104,745	101,225	Kapuskasing	92,214	88,830
Summerland	118,093	113,833	Kingston	134,085	129,099
Terrace	82,773	80,033	Lindsay	117,889	113,655
Trail	95,446	92,092	London	118,827	114,078
Vancouver	116,222	111,960	Mount Forest	119,642	115,073
Victoria	134,579	129,792	New Liskeard	96,711	93,369
MANITOBA			North Bay	120,541	116,246
Brandon	129,530	125,375	Ottawa	125,000	120,673
Dauphin	134,070	129,714	St. Catharines	113,609	109,044
Morden	135,418	130,734	Sarnia	127,619	122,667
Rivers	139,439	134,903	Sault Ste. Marie	116,534	112,300
The Pas	111,980	108,455	Thunder Bay	130,984	126,508
Winnipeg	138,267	133,879	Toronto	132,387	127,688
NEW BRUNSWICK			Woodstock	121,200	116,512
Campbellton	121,858	117,680	PRINCE EDWARD ISLAND		
Chatham	128,564	124,405	Charlottetown	114,927	111,233
Fredericton	123,862	119,842	QUEBEC		
Moncton	125,591	121,411	Amos	103,142	99,356
Saint John	92,117	89,144	Berthierville	132,336	127,888

	Btu Output for 60°-Tilt Collector	Btu Output for 90°-Tilt Collector		Btu Output for 60°-Tilt Collector	Btu Output for 90°-Tilt Collector
Grandes Bergeronnes	116,649	112,760	SASKATCHEWAN		
L'Assomption	127,779	123,380	Estevan	155,448	150,205
Lennoxville	113,363	109,441	Indian Head	139,078	134,600
Maniwaki	121,149	116,680	Melfort	115,721	112,186
Montreal	134,559	130,030	Moose Jaw	138,929	134,341
Normandin	109,018	105,166	Outlook	131,611	127,222
Quebec	110,777	107,109	Prince Albert	120,900	117,162
Rimouski	96,152	92,820	Regina	134,565	130,000
Ste. Agathe-des-Monts	124,421	119,888	Saskatoon	141,260	136,795
St. Ambroise	107,191	103,277	Scott	121,915	117,850
Sept-Iles	126,074	121,612	Swift Current	135,879	131,375
Sherbrooke	118,516	114,447	Weyburn	144,392	139,771
Victoriaville	123,291	119,116	Yorkton	125,793	121,751

Note to Table 47: *Represents 273-day heating season total Btu output from 1 square foot of collector derived from Tables 50 and 51 (Average Daily Btu Output by Month From 1 Square Foot of Properly Designed Collector). See Notes following Table 51 for conditions.*

SECTION SEVEN

Ranking of North American Cities by Size of Solar Furnace Needed

15 The Practicality of Solar Energy in North America

THE MYTH OF THE "SUN BELT"

One of the most surprising things about the practical utilization of solar energy for heating homes is the fact that it is not confined to some optimum "sun belt" areas. The very worst town for using solar energy, in far northern Manitoba, still receives sufficient sunshine to deliver 40% of an average town's Btu delivery inside the home.

CLOUDINESS

The things that conspire to make a solar furnace work less efficiently are not confined to the amount of cloud cover. We have seen in earlier chapters that the properly designed, properly insulated, low-temperature solar furnace will collect energy even on a cloudy day, albeit not so efficiently.

If you doubt this, just think back to that bad sunburn you got at the beach on a cloudy day. Solar energy penetrates clouds.

LATITUDE

The farther north you live, the farther the sunshine must travel through the atmosphere to reach you, and the more oblique is its angle.

TEMPERATURE

The colder the temperature outside, the larger the difference between the temperature inside the collector and inside the storage unit and outdoors, which means that the heat loss from the collector increases. That means that energy is lost through the covers, sides and back of the collector instead of being transferred to storage, and the losses from storage increase, as well.

TRANSMISSIVITY OF THE COVERS

Although theoretically the optimum tilt of the collector for heating in the winter is latitude plus 20°, the realities of mass production mean that most of the units sold on the North American continent so far (some 90%, as of this writing) have been self-contained solar furnaces combining collector, fans and storage in one ground-mounted unit. The manufacturers of this type of unit have adopted a uniform collector tilt of 60° or 90° on all units made.

This means that, if you live in a city at Latitude 40 degrees North, the 60° tilt is ideal. But if you live in northern Canada at Latitude 55 degrees North, that collector is about 15° away from optimum tilt.

This means that losses due to reflectivity increase, and slightly less energy strikes each square foot of the collector than would in midwinter if the collector had a tilt of 75°.

It turns out that this ins't a big deal, in the real world. Later on during the heating season, the optimum collector tilt is latitude plus 5°; so people living in northern Canada purchasing mass-produced solar units with a 60° tilt simply have better collection later on in the heating season.

For example, during April and May a solar furnace having either a 60° tilt or a 90° tilt and situated in Fort St. John, British Columbia (Latitude 56 degrees North), collects more energy than one situated in Elkins, West Virginia (Latitude 39 degrees North). And as it turns out, that far north the heating demands are still high during those months; so the mass-produced units with their compromise collector tilts work out very well.

PUTTING IT ALL TOGETHER

As you have already seen if you performed any of the calculations in Chapter Thirteen, it is necessary to take into account all the things just discussed—cloudiness, latitude, transmissivity of covers, amount of insulation on both collector and storage, the number of collector covers and the transmissivity of the covers—in order to determine the output of a solar furnace in a given community. Then one can determine the size of solar furnace required for a given house.

WHAT THE RANKING MEANS

Taking the very same size house, with the same insulation, with the same percentage of solar heat supplied, and "placing" it in each of 342 cities, resulted in the lists that follow. The cities are listed in ascending order according to the size of solar furnace

required there (Table 49). This ranking then shows the relative merit of those communities from a practical solar energy utilization standpoint (see Table 48, listing States and Provinces in alphabetical order).

SURPRISES

You will, in glancing at the lists that follow, soon see that one should not make assumptions concerning which region, state or province is best for solar utilization.

For example, both San Francisco, California, and Washington, D.C., rank higher than purportedly ideal Denver, Colorado. The so-called Cloud Capital of the United States, Seattle, Washington, ranks higher than Lexington, Kentucky. Wintry St. John, New Brunswick, ranks higher than Dubuque, Iowa. Similarly, New York City and Trenton, New Jersey, are better for solar utilization than Grand Junction, Colorado, and Elko, Nevada.

It is to be hoped that this proves one thing about the practical utilization of solar energy: You simply should not make assumptions. To determine its practicality, you must perform the necessary calculations.

The only way you can see if a solar furnace is a good idea for your home is to check the Btu output for your community, see what you are paying for fuel now, and check the pay-back time. And that will require knowing just how large, in Degree Days, your house is, how much solar furnace it will require, and how much that is going to cost you. That takes time—probably an hour or two. But it is the only way to find out. There are no easy ways, using rules of thumb.

Two lists follow. One gives the numerical listing, in descending order, of the cities. The other is arranged by State and Province, so that you can more easily find your home town.

Table 48 342 NORTH AMERICAN CITIES, RANKED IN ASCENDING ORDER OF SIZE OF SOLAR FURNACE REQUIRED, LISTED BY STATE OR PROVINCE

STATE/ Town	Rank by Size of Solar Furnace Needed	STATE/ Town	Rank by Size of Solar Furnace Needed	STATE/ Town	Rank by Size of Solar Furnace Needed
ALABAMA		CONNECTICUT		Rockford	198
Birmingham	70	Bridgeport	144	Springfield	147
Huntsville	85	Hartford	171	INDIANA	
Mobile	29	New Haven	128	Evansville	121
Montgomery	51	DELAWARE		Fort Wayne	189
ARIZONA		Wilmington	114	Indianapolis	169
Flagstaff	102	DISTRICT OF COLUMBIA		South Bend	206
Phoenix	15	Washington	99	IOWA	
Prescott	67	FLORIDA		Burlington	156
Tucson	14	Appalachicola	17	Des Moines	182
Winslow	79	Daytona Beach	6	Dubuque	204
Yuma	7	Jacksonville	19	Sioux City	181
ARKANSAS		Key West	1	Waterloo	193
Fort Smith	94	Miami Beach	2	KANSAS	
Little Rock	90	Orlando	3	Concordia	110
Texarkana	63	Pensacola	26	Dodge City	92
CALIFORNIA		Tallahassee	23	Goodland	108
Bakersfield	32	Tampa	4	Topeka	124
Bishop	73	GEORGIA		Wichita	96
Burbank	11	Athens	68	KENTUCKY	
Eureka	89	Atlanta	72	Covington	155
Fresno	44	Augusta	54	Lexington	145
Long Beach	10	Columbus	50	Louisville	137
Los Angeles	13	Macon	37	LOUISIANA	
Oakland	39	Rome	80	Alexandria	38
Red Bluff	53	Savannah	25	Baton Rouge	27
Sacramento	59	IDAHO		Lake Charles	21
San Diego	8	Boise	150	New Orleans	20
San Francisco	40	Idaho Falls	239	Shreveport	45
Santa Maria	30	Lewiston	194	MAINE	
COLORADO		Pocatello	210	Caribou	304
Alamosa	186	ILLINOIS		Portland	165
Colorado Springs	101	Cairo	91	MARYLAND	
Denver	107	Chicago	184	Baltimore	106
Grand Junction	117	Moline	191	Frederick	130
Pueblo	86	Peoria	170	MASSACHUSETTS	

STATE/ Town	Rank by Size of Solar Furnace Needed	STATE/ Town	Rank by Size of Solar Furnace Needed	STATE/ Town	Rank by Size of Solar Furnace Needed
Boston	140	Grand Island	153	Devils Lake	164
Nantucket	132	Lincoln	134	Fargo	157
Pittsfield	222	Norfolk	173	Williston	151
Worcester	187	North Platte	136	OHIO	
MICHIGAN		Omaha	162	Akron	215
Alpena	250	Scottsbluff	127	Cincinnati	118
Detroit	212	NEVADA		Cleveland	218
Escanaba	231	Elko	143	Columbus	180
Flint	227	Ely	131	Dayton	146
Grand Rapids	225	Las Vegas	34	Mansfield	207
Lansing	220	Reno	97	Sandusky	190
Marquette	257	Winnemucca	115	Toledo	119
Muskegon	216	NEW HAMPSHIRE		Youngstown	133
Sault Ste. Marie	275	Concord	203	OKLAHOMA	
MINNESOTA		NEW JERSEY		Oklahoma City	74
Duluth	248	Atlantic City	100	Tulsa	83
International Falls	283	Newark	112	OREGON	
Minneapolis	217	Trenton	116	Astoria	161
Rochester	213	NEW MEXICO		Eugene	159
Saint Cloud	238	Albuquerque	62	Medford	160
MISSISSIPPI		Raton	98	Pendleton	185
Jackson	61	Roswell	56	Portland	163
Meridian	57	Silver City	55	Roseburg	158
Vicksburg	47	NEW YORK		Salem	172
MISSOURI		Albany	201	PENNSYLVANIA	
Columbia	123	Binghampton	230	Allentown	135
Kansas City	104	Buffalo	229	Erie	219
St. Joseph	138	New York City	111	Harrisburg	120
St. Louis	122	Rochester	223	Philadelphia	109
Springfield	105	Schenectady	200	Pittsburgh	205
MONTANA		Syracuse	234	Reading	125
Billings	177	NORTH CAROLINA		Scranton	188
Glasgow	228	Asheville	84	Williamsport	195
Great Falls	174	Charlotte	65	RHODE ISLAND	
Havre	224	Greensboro	77	Providence	126
Helena	202	Raleigh	69	SOUTH CAROLINA	
Kalispell	241	Wilmington	46	Charleston	33
Miles City	192	Winston-Salem	93	Columbia	60
Missoula	240	NORTH DAKOTA		Florence	52
NEBRASKA		Bismark	142	Greenville	71

STATE/ Town	Rank by Size of Solar Furnace Needed	STATE/ Town	Rank by Size of Solar Furnace Needed	STATE/ Town	Rank by Size of Solar Furnace Needed
Spartanburg	75	WASHINGTON		Kamloops	318
SOUTH DAKOTA		Olympia	167	Kitimat	325
Huron	208	Seattle	141	Nanaimo	236
Rapid City	148	Spokane	214	Nelson	310
Sioux Falls	196	Tacoma	176	New Westminster	243
TENNESSEE		Walla Walla	178	Penticton	287
Bristol	103	Yakima	166	Port Alberni	251
Chattanooga	81	WEST VIRGINIA		Powell River	246
Knoxville	87	Charleston	168	Prince George	331
Memphis	78	Elkins	179	Prince Rupert	312
Nashville	95	Huntington	152	Quesnel	322
TEXAS		Parkersburg	183	Summerland	303
Abilene	43	WISCONSIN		Terrace	330
Amarillo	58	Green Bay	232	Trail	334
Austin	42	La Crosse	221	Vancouver	237
Brownsville	5	Madison	226	Victoria	199
Corpus Christi	12	Milwaukee	209	MANITOBA	
Dallas	48	WYOMING		Brandon	316
El Paso	35	Casper	149	Dauphin	306
Fort Worth	49	Cheyenne	129	Morden	299
Galveston	18	Lander	154	Rivers	300
Houston	22	Sheridan	175	The Pas	342
Laredo	9	ALBERTA		Winnipeg	311
Lubbock	64	Banff	324	NEW BRUNSWICK	
Midland	36	Beaverlodge	321	Campbellton	279
Port Arthur	24	Brooks	281	Chatham	259
San Angelo	31	Calgary	265	Fredericton	233
San Antonio	28	Edmonton	282	Moncton	252
Victoria	16	Edson	307	Saint John	197
Waco	41	Fairview	340	NEWFOUNDLAND	
Wichita Falls	66	Lacombe	288	Gander	317
UTAH		Lethbridge	249	St. John's	296
Salt Lake City	139	Medicine Hat	273	NOVA SCOTIA	
VERMONT		Olds	276	Annapolis Royal	262
Burlington	258	Vauxhall	264	Halifax	211
VIRGINIA		BRITISH COLUMBIA		Kentville	266
Lynchburg	88	Campbell River	244	Sydney	255
Norfolk	76	Dawson Creek	332	Truro	260
Richmond	82	Fort St. John	338	Yarmouth	235
Roanoke	113	Hope	294	ONTARIO	

STATE/ Town	Rank by Size of Solar Furnace Needed	STATE/ Town	Rank by Size of Solar Furnace Needed	STATE/ Town	Rank by Size of Solar Furnace Needed
Atikokan	315	Sault Ste. Marie	295	St. Ambroise	327
Belleville	267	Thunder Bay	284	Sept-Iles	286
Brampton	254	Toronto	245	Sherbrooke	290
Chatham	256	Woodstock	274	Victoriaville	291
Cornwall	271	PRINCE EDWARD ISLAND		SASKATCHEWAN	
Guelph	270	Charlottetown	269	Estevan	268
Hamilton	242	QUEBEC		Indian Head	314
Harrow	261	Amos	335	Melfort	337
Kapuskasing	341	Berthierville	278	Moose Jaw	309
Kingston	247	Grandes Bergeronnes	298	Outlook	320
Lindsay	289	L'Assomption	285	Prince Albert	339
London	280	Lennoxville	308	Regina	323
Mount Forest	293	Maniwaki	305	Saskatoon	319
New Liskeard	333	Montreal	263	Scott	336
North Bay	292	Normandin	329	Swift Current	302
Ottawa	277	Quebec	313	Weyburn	297
St. Catharines	272	Rimouski	326	Yorkton	328
Sarnia	253	Ste. Agathe-des-Monts	301		

Table 49 RANKING OF 342 NORTH AMERICAN CITIES BY SIZE OF SOLAR FURNACE REQUIRED, IN ASCENDING ORDER

1. Key West, Florida
2. Miami Beach, Florida
3. Orlando, Florida
4. Tampa, Florida
5. Brownsville, Texas
6. Daytona Beach, Florida
7. Yuma, Arizona
8. San Diego, California
9. Laredo, Texas
10. Long Beach, California
11. Burbank, California
12. Corpus Christi, Texas
13. Los Angeles, California
14. Tucson, Arizona
15. Phoenix, Arizona
16. Victoria, Texas
17. Appalachicola, Florida
18. Galveston, Texas
19. Jacksonville, Florida
20. New Orleans, Louisiana
21. Lake Charles, Louisiana
22. Houston, Texas
23. Tallahassee, Florida
24. Port Arthur, Texas
25. Savannah, Georgia
26. Pensacola, Florida
27. Baton Rouge, Louisiana
28. San Antonio, Texas
29. Mobile, Alabama
30. Santa Maria, California
31. San Angelo, Texas
32. Bakersfield, California
33. Charleston, South Carolina
34. Las Vegas, Nevada
35. El Paso, Texas
36. Midland, Texas
37. Macon, Georgia
38. Alexandria, Louisiana
39. Oakland, California
40. San Francisco, California
41. Waco, Texas

42. Austin, Texas
43. Abilene, Texas
44. Fresno, California
45. Shreveport, Louisiana
46. Wilmington, North Carolina
47. Vicksburg, Mississippi
48. Dallas, Texas
49. Fort Worth, Texas
50. Columbus, Georgia
51. Montgomery, Alabama
52. Florence, South Carolina
53. Red Bluff, California
54. Augusta, Georgia
55. Silver City, New Mexico
56. Roswell, New Mexico
57. Meridian, Mississippi
58. Amarillo, Texas
59. Sacramento, California
60. Columbia, South Carolina
61. Jackson, Mississippi
62. Albuquerque, New Mexico
63. Texarkana, Arkansas
64. Lubbock, Texas
65. Charlotte, North Carolina
66. Wichita Falls, Texas
67. Prescott, Arizona
68. Athens, Georgia
69. Raleigh, North Carolina
70. Birmingham, Alabama
71. Greenville, South Carolina
72. Atlanta, Georgia
73. Bishop, California
74. Oklahoma City, Oklahoma
75. Spartanburg, South Carolina
76. Norfolk, Virginia
77. Greensboro, North Carolina
78. Memphis, Tennessee
79. Winslow, Arizona
80. Rome, Georgia
81. Chattanooga, Tennessee
82. Richmond, Virginia

83. Tulsa, Oklahoma
84. Asheville, North Carolina
85. Huntsville, Alabama
86. Pueblo, Colorado
87. Knoxville, Tennessee
88. Lynchburg, Virginia
89. Eureka, California
90. Little Rock, Arkansas
91. Cairo, Illinois
92. Dodge City, Kansas
93. Winston-Salem, North Carolina
94. Fort Smith, Arkansas
95. Nashville, Tennessee
96. Wichita, Kansas
97. Reno, Nevada
98. Raton, New Mexico
99. Washington, D.C.
100. Atlantic City, New Jersey
101. Colorado Springs, Colorado
102. Flagstaff, Arizona
103. Bristol, Tennessee
104. Kansas City, Missouri
105. Springfield, Missouri
106. Baltimore, Maryland
107. Denver, Colorado
108. Goodland, Kansas
109. Philadelphia, Pennsylvania
110. Concordia, Kansas
111. New York City, New York
112. Newark, New Jersey
113. Roanoke, Virginia
114. Wilmington, Delaware
115. Winnemucca, Nevada
116. Trenton, New Jersey
117. Grand Junction, Colorado
118. Cincinnati, Ohio
119. Toledo, Ohio
120. Harrisburg, Pennsylvania
121. Evansville, Indiana
122. St. Louis, Missouri

123. Columbia, Missouri
124. Topeka, Kansas
125. Reading, Pennsylvania
126. Providence, Rhode Island
127. Scottsbluff, Nebraska
128. New Haven, Connecticut
129. Cheyenne, Wyoming
130. Frederick, Maryland
131. Ely, Nevada
132. Nantucket, Massachusetts
133. Youngstown, Ohio
134. Lincoln, Nebraska
135. Allentown, Pennsylvania
136. North Platte, Nebraska
137. Louisville, Kentucky
138. St. Joseph, Missouri
139. Salt Lake City, Utah
140. Boston, Massachusetts
141. Seattle, Washington
142. Bismarck, North Dakota
143. Elko, Nevada
144. Bridgeport, Connecticut
145. Lexington, Kentucky
146. Dayton, Ohio
147. Springfield, Illinois
148. Rapid City, South Dakota
149. Casper, Wyoming
150. Boise, Idaho
151. Williston, North Dakota
152. Huntington, West Virginia
153. Grand Island, Nebraska
154. Lander, Wyoming
155. Covington, Kentucky
156. Burlington, Iowa
157. Fargo, North Dakota
158. Roseburg, Oregon
159. Eugene, Oregon
160. Medford, Oregon
161. Astoria, Oregon
162. Omaha, Nebraska
163. Portland, Oregon
164. Devils Lake, North Dakota
165. Portland, Maine
166. Yakima, Washington

167. Olympia, Washington
168. Charleston, West Virginia
169. Indianapolis, Indiana
170. Peoria, Illinois
171. Hartford, Connecticut
172. Salem, Oregon
173. Norfolk, Nebraska
174. Great Falls, Montana
175. Sheridan, Wyoming
176. Tacoma, Washington
177. Billings, Montana
178. Walla Walla, Washington
179. Elkins, West Virginia
180. Columbus, Ohio
181. Sioux City, Iowa
182. Des Moines, Iowa
183. Parkersburg, West Virginia
184. Chicago, Illinois
185. Pendleton, Oregon
186. Alamosa, Colorado
187. Worcester, Massachusetts
188. Scranton, Pennsylvania
189. Fort Wayne, Indiana
190. Sandusky, Ohio
191. Moline, Illinois
192. Miles City, Montana
193. Waterloo, Iowa
194. Lewiston, Idaho
195. Williamsport, Pennsylvania
196. Sioux Falls, South Dakota
197. Saint John, New Brunswick
198. Rockford, Illinois
199. Victoria, British Columbia
200. Schenectady, New York
201. Albany, New York
202. Helena, Montana
203. Concord, New Hampshire
204. Dubuque, Iowa
205. Pittsburgh, Pennsylvania
206. South Bend, Indiana
207. Mansfield, Ohio
208. Huron, South Dakota
209. Milwaukee, Wisconsin
210. Pocatello, Idaho

211. Halifax, Nova Scotia
212. Detroit, Michigan
213. Rochester, Minnesota
214. Spokane, Washington
215. Akron, Ohio
216. Muskegon, Michigan
217. Minneapolis, Minnesota
218. Cleveland, Ohio
219. Erie, Pennsylvania
220. Lansing, Michigan
221. La Crosse, Wisconsin
222. Pittsfield, Massachusetts
223. Rochester, New York
224. Havre, Montana
225. Grand Rapids, Michigan
226. Madison, Wisconsin
227. Flint, Michigan
228. Glasgow, Montana
229. Buffalo, New York
230. Binghampton, New York
231. Escanaba, Michigan
232. Green Bay, Wisconsin
233. Fredericton, New Brunswick
234. Syracuse, New York
235. Yarmouth, Nova Scotia
236. Nanaimo, British Columbia
237. Vancouver, British Columbia
238. St. Cloud, Minnesota
239. Idaho Falls, Idaho
240. Missoula, Montana
241. Kalispell, Montana
242. Hamilton, Ontario
243. New Westminster, British Columbia
244. Campbell River, British Columbia
245. Toronto, Ontario
246. Powell River, British Columbia
247. Kingston, Ontario
248. Duluth, Minnesota
249. Lethbridge, Alberta
250. Alpena, Michigan
251. Port Alberni, British

Columbia
252. Moncton, New Brunswick
253. Sarnia, Ontario
254. Brampton, Ontario
255. Sydney, Nova Scotia
256. Chatham, Ontario
257. Marquette, Michigan
258. Burlington, Vermont
259. Chatham, New Brunswick
260. Truro, Nova Scotia
261. Harrow, Ontario
262. Annapolis Royal, Nova Scotia
263. Montreal, Quebec
264. Vauxhall, Alberta
265. Calgary, Alberta
266. Kentville, Nova Scotia
267. Belleville, Ontario
268. Estevan, Saskatchewan
269. Charlottetown, Prince Edward Island
270. Guelph, Ontario
271. Cornwall, Ontario
272. St. Catharines, Ontario
273. Medicine Hat, Alberta
274. Woodstock, Ontario
275. Sault Ste. Marie, Michigan
276. Olds, Alberta
277. Ottawa, Ontario
278. Berthierville, Quebec
279. Campbellton, New Brunswick
280. London, Ontario
281. Brooks, Alberta
282. Edmonton, Alberta

283. International Falls, Minnesota
284. Thunder Bay, Ontario
285. L'Assomption, Quebec
286. Sept-Iles, Quebec
287. Penticton, British Columbia
288. Lacombe, Alberta
289. Lindsay, Ontario
290. Sherbrooke, Quebec
291. Victoriaville, Quebec
292. North Bay, Ontario
293. Mount Forest, Ontario
294. Hope, British Columbia
295. Sault Ste. Marie, Ontario
296. St. John's, Newfoundland
297. Weyburn, Saskatchewan
298. Grandes Bergeronnes, Quebec
299. Morden, Manitoba
300. Rivers, Manitoba
301. Ste. Agathe-des-Monts, Quebec
302. Swift Current, Saskatchewan
303. Summerland, British Columbia
304. Caribou, Maine
305. Maniwaki, Quebec
306. Dauphin, Manitoba
307. Edson, Alberta
308. Lennoxville, Quebec
309. Moose Jaw, Saskatchewan
310. Nelson, British Columbia
311. Winnipeg, Manitoba

312. Prince Rupert, British Columbia
313. Quebec, Quebec
314. Indian Head, Saskatchewan
315. Atikokan, Ontario
316. Brandon, Manitoba
317. Gander, Newfoundland
318. Kamloops, British Columbia
319. Saskatoon, Saskatchewan
320. Outlook, Saskatchewan
321. Beaverlodge, Alberta
322. Quesnel, British Columbia
323. Regina, Saskatchewan
324. Banff, Alberta
325. Kitimat, British Columbia
326. Rimouski, Quebec
327. St. Ambroise, Quebec
328. Yorkton, Saskatchewan
329. Normandin, Quebec
330. Terrace, British Columbia
331. Prince George, British Columbia
332. Dawson Creek, British Columbia
333. New Liskeard, Ontario
334. Trail, British Columbia
335. Amos, Quebec
336. Scott, Saskatchewan
337. Melfort, Saskatchewan
338. Fort St. John, British Columbia
339. Prince Albert, Saskatchewan
340. Fairview, Alberta
341. Kapuskasing, Ontario
342. The Pas, Manitoba

Notes to Tables 48 and 49: *Derived from calculated percentages of supply of solar furnace on same DD home in 342 cities using procedure described in Appendix D.*

SECTION EIGHT

Appendices

Appendix A

Table 50 AVERAGE DAILY B.T.U. OUTPUT INSIDE THE HOUSE FROM 60°-TILT DOUBLE-COVER COLLECTOR WITH HORIZONTAL REFLECTOR, PER SQUARE FOOT

	Sept	Oct	Nov	Dec	Jan	Feb	Mar	Apr	May
ALABAMA									
Birmingham	819	962	887	670	658	705	736	716	597
Huntsville	814	981	857	613	633	695	728	714	599
Mobile	762	1015	947	734	777	747	750	708	603
Montgomery	842	1007	973	736	780	782	790	762	634
ARIZONA									
Flagstaff	979	1035	970	930	803	925	850	775	704
Phoenix	1105	1263	1316	1216	1162	1180	1118	1000	841
Prescott	1065	1156	1212	1067	1018	1045	970	901	784
Tucson	1108	1284	1390	1316	1238	1276	1149	1009	824
Winslow	1052	1170	1153	1011	910	1018	983	934	793
Yuma	1145	1327	1439	1316	1316	1321	1245	1053	859
ARKANSAS									
Fort Smith	877	953	828	701	674	738	702	691	572
Little Rock	889	1068	857	687	631	767	755	714	618
Texarkana	906	1091	948	807	736	792	776	716	624
CALIFORNIA									
Bakersfield	1128	1227	1225	835	841	1052	1017	922	784
Bishop	1094	1001	853	754	915	907	908	792	770
Burbank	1030	1062	1175	1106	1111	1101	1010	795	588
Eureka	666	687	624	521	568	642	670	631	564
Fresno	1183	1274	1114	687	691	947	969	962	877
Long Beach	1030	1062	1175	1106	1111	1101	1010	777	588
Los Angeles	993	1091	1238	1165	1111	1030	942	744	601
Oakland	1025	1061	913	718	744	873	949	846	775
Red Bluff	1158	1151	968	690	702	904	896	906	853
Sacramento	1188	1217	986	619	624	859	916	906	860
San Diego	862	999	1215	1153	1079	1018	909	740	531
San Francisco	897	1032	943	775	788	858	854	811	696
Santa Maria	977	1054	1115	1013	979	1082	927	765	586
COLORADO									
Alamosa	939	783	796	639	736	762	692	585	582
Colorado Springs	930	955	891	899	902	927	865	695	610
Denver	923	1012	916	819	855	888	809	716	643
Grand Junction	994	1048	919	714	737	821	812	780	744
Pueblo	1000	1069	1017	930	963	980	870	780	676

	Sept	Oct	Nov	Dec	Jan	Feb	Mar	Apr	May
CONNECTICUT									
Bridgeport	812	787	717	609	622	679	653	620	663
Hartford	753	791	623	519	548	705	709	626	619
New Haven	825	888	730	658	646	799	728	677	653
DELAWARE									
Wilmington	806	891	773	617	599	763	729	684	659
DISTRICT OF COLUMBIA									
Washington D.C.	800	885	790	620	635	721	710	679	639
FLORIDA									
Appalachicola	743	1038	1022	849	935	924	811	762	640
Daytona Beach	652	809	978	982	1056	948	875	749	560
Jacksonville	701	818	970	846	919	885	869	773	603
Key West	735	886	1070	1096	1109	1109	998	761	557
Miami Beach	709	850	1012	1081	1080	1071	942	727	512
Orlando	687	837	978	1014	1073	978	915	759	584
Pensacola	810	1058	1009	846	872	870	829	762	620
Tallahassee	731	982	1007	897	888	895	850	752	607
Tampa	749	929	1053	996	1015	1010	933	765	594
GEORGIA									
Athens	857	1005	948	733	751	777	802	733	633
Atlanta	807	962	918	689	707	763	749	722	615
Augusta	862	1027	931	795	795	801	809	784	629
Columbus	776	1027	946	764	795	816	809	751	647
Macon	837	956	1000	825	856	860	855	807	664
Rome	782	912	855	645	663	720	697	688	606
Savannah	695	866	965	859	872	826	823	773	625
IDAHO									
Boise	1087	996	584	405	458	664	779	834	758
Idaho Falls	983	806	498	307	361	535	624	682	664
Lewiston	967	872	398	192	310	536	670	774	709
Pocatello	788	711	513	364	405	527	611	646	598
ILLINOIS									
Cairo	954	1044	823	631	648	739	765	753	699
Chicago	858	877	620	463	525	655	671	649	684
Moline	885	891	580	486	537	602	628	626	608
Peoria	904	916	685	502	564	666	678	679	642
Rockford	872	877	593	467	525	577	665	638	662
Springfield	949	935	744	567	600	676	672	684	691
INDIANA									
Evansville	936	965	734	565	570	667	692	717	681
Fort Wayne	839	830	550	444	472	586	640	631	663
Indianapolis	884	913	674	471	523	623	610	649	654

Table 50   189

	Sept	Oct	Nov	Dec	Jan	Feb	Mar	Apr	May
South Bend	845	848	501	395	441	525	641	626	684
IOWA									
Burlington	943	1002	792	596	613	732	741	715	674
Des Moines	845	920	699	534	639	718	690	684	673
Dubuque	787	762	565	445	547	657	641	672	651
Sioux City	885	920	699	534	639	756	715	719	737
Waterloo	879	894	604	483	577	698	707	679	682
KANSAS									
Concordia	936	1022	898	730	766	795	772	743	702
Dodge City	974	1080	988	877	872	898	856	620	692
Goodland	942	1005	905	813	876	914	840	740	686
Topeka	904	957	802	655	683	721	659	628	619
Wichita	936	994	946	773	828	857	805	752	671
KENTUCKY									
Covington	878	870	647	501	551	626	642	652	650
Lexington	833	850	677	471	516	585	617	631	610
Louisville	871	922	720	525	556	639	654	654	651
LOUISIANA									
Alexandria	871	1058	978	798	718	762	737	730	578
Baton Rouge	834	1015	947	766	777	750	750	664	569
Lake Charles	839	1052	960	769	761	775	719	666	574
New Orleans	767	982	929	736	777	745	745	676	549
Shreveport	973	1099	1015	917	734	801	757	672	611
MAINE									
Caribou	550	567	318	240	291	478	496	501	490
Portland	800	824	635	559	609	762	726	637	642
MARYLAND									
Baltimore	826	885	790	629	659	762	723	675	639
Frederick	800	850	704	578	564	694	659	605	639
MASSACHUSETTS									
Boston	806	690	650	565	569	749	722	649	640
Nantucket	852	859	716	563	557	692	728	630	648
Pittsfield	650	695	492	333	476	643	633	575	580
Worcester	748	748	580	497	513	679	665	591	604
MICHIGAN									
Alpena	686	617	291	219	304	550	620	626	651
Detroit	806	776	474	327	411	538	591	603	630
Escanaba	692	673	375	334	403	644	660	632	613
Flint	754	695	388	290	371	630	607	597	623
Grand Rapids	772	721	401	316	325	507	596	632	660
Lansing	785	721	414	305	371	528	633	632	671
Marquette	630	533	272	219	300	499	577	592	577

	Sept	Oct	Nov	Dec	Jan	Feb	Mar	Apr	May
Muskegon	785	692	388	305	337	502	633	644	634
Sault Ste. Marie	604	505	238	192	267	525	605	615	588
MINNESOTA									
Duluth	694	634	392	343	427	647	694	660	632
International Falls	718	648	340	304	369	597	634	661	670
Minneapolis	792	757	485	381	505	660	665	654	678
Rochester	826	852	519	406	531	658	648	648	664
Saint Cloud	785	771	427	377	464	653	636	672	659
MISSISSIPPI									
Jackson	768	894	897	736	627	693	700	696	573
Meridian	829	951	897	720	749	737	751	718	599
Vicksburg	903	1007	934	690	703	737	756	707	599
MISSOURI									
Columbia	904	943	774	604	631	735	710	675	681
Kansas City	904	972	830	668	739	776	748	715	671
St. Joseph	910	949	786	630	664	729	710	684	659
St. Louis	865	943	760	565	645	667	710	687	671
Springfield	903	930	821	658	658	734	711	695	620
MONTANA									
Billings	905	883	579	464	526	679	724	698	697
Glasgow	878	809	505	382	436	684	741	732	720
Great Falls	918	855	515	449	553	763	828	755	720
Havre	851	799	487	363	428	676	759	750	728
Helena	845	769	545	392	445	696	742	701	666
Kalispell	805	675	307	174	260	500	631	673	662
Miles City	885	883	590	465	495	761	749	765	740
Missoula	851	715	363	228	300	481	576	596	633
NEBRASKA									
Grand Island	930	959	792	643	703	792	758	712	696
Lincoln	878	945	813	643	699	786	753	715	674
Norfolk	898	920	725	556	650	769	739	707	695
North Platte	943	1002	832	678	772	839	782	712	685
Omaha	943	1031	752	596	680	804	733	746	728
Scottsbluff	924	942	791	689	751	882	801	719	673
NEVADA									
Elko	1062	994	784	608	686	799	758	696	700
Ely	1022	996	894	763	775	813	818	702	646
Las Vegas	1161	1221	1270	1076	1080	1116	1041	946	810
Reno	1093	1045	930	705	753	871	882	853	793
Winnemucca	1101	1035	832	619	646	799	804	804	794
NEW HAMPSHIRE									
Concord	715	695	556	462	533	682	683	597	561

Table 50   191

	Sept	Oct	Nov	Dec	Jan	Feb	Mar	Apr	May
NEW JERSEY									
Atlantic City	845	789	846	673	679	776	741	696	669
Newark	839	894	759	609	635	766	735	703	679
Trenton	832	876	773	617	639	763	685	696	669
NEW MEXICO									
Albuquerque	1015	1128	1124	996	977	1060	1043	1030	910
Raton	916	979	965	827	920	953	851	709	611
Roswell	911	1032	1088	988	989	991	943	843	673
Silver City	1046	1129	1191	1103	1032	1101	1031	898	717
NEW YORK									
Albany	772	779	505	414	499	656	658	620	627
Binghampton	621	618	393	294	375	500	505	510	543
Buffalo	799	736	401	305	371	528	608	575	634
New York City	839	894	759	609	653	766	716	703	679
Rochester	799	721	438	338	371	540	583	609	660
Schenectady	772	779	505	414	499	656	658	620	627
Syracuse	745	678	385	283	360	489	558	585	638
NORTH CAROLINA									
Asheville	777	902	872	683	688	739	715	686	590
Charlotte	827	996	982	803	780	840	791	765	664
Greensboro	821	952	899	755	724	781	759	674	648
Raleigh	795	909	914	756	730	781	759	731	639
Wilmington	795	976	1034	883	872	835	841	807	660
Winston-Salem	795	753	855	732	682	725	720	708	620
NORTH DAKOTA									
Bismark	832	827	533	418	472	692	677	666	655
Devils Lake	797	755	463	368	455	697	719	690	674
Fargo	805	799	425	395	427	647	677	678	700
Williston	878	809	505	392	445	696	731	744	754
OHIO									
Akron	812	787	455	292	361	479	565	597	663
Cincinnati	878	870	647	501	551	626	659	652	650
Cleveland	812	773	441	292	361	479	550	597	653
Columbus	858	856	618	441	460	583	610	637	665
Dayton	936	941	716	541	485	597	660	660	665
Mansfield	839	802	510	350	410	546	587	597	663
Sandusky	872	834	514	377	411	575	633	661	684
Toledo	819	791	501	339	382	551	604	614	651
Youngstown	799	745	430	292	348	519	550	574	642
OKLAHOMA									
Oklahoma City	927	959	946	811	817	836	804	720	599
Tulsa	947	980	988	838	696	753	695	616	534

	Sept	Oct	Nov	Dec	Jan	Feb	Mar	Apr	May
OREGON									
Astoria	599	596	347	234	309	451	536	578	536
Eugene	859	608	379	231	294	469	548	613	558
Medford	872	719	514	302	353	550	624	625	597
Pendleton	1008	872	470	218	390	522	670	762	812
Portland	749	611	360	245	309	479	571	612	594
Roseburg	905	606	386	216	303	453	544	619	627
Salem	811	611	372	223	309	471	532	619	565
PENNSYLVANIA									
Allentown	812	845	675	560	547	719	703	654	653
Erie	806	704	393	282	339	500	567	568	630
Harrisburg	806	847	660	541	573	689	685	672	659
Philadelphia	806	891	744	617	599	763	729	684	659
Pittsburgh	812	773	538	350	398	519	564	574	610
Reading	767	832	700	554	586	554	677	649	626
Scranton	747	759	537	432	448	612	603	574	599
Williamsport	721	716	482	362	423	559	603	563	578
RHODE ISLAND									
Providence	819	863	677	518	581	775	684	638	651
SOUTH CAROLINA									
Charleston	825	970	1062	871	887	890	869	807	647
Columbia	795	976	979	776	811	820	815	773	624
Florence	782	962	964	761	795	806	802	761	615
Greenville	802	953	951	745	780	825	808	806	655
Spartanburg	777	924	905	701	707	767	715	703	590
SOUTH DAKOTA									
Huron	871	855	630	467	602	758	726	712	735
Rapid City	904	921	736	567	654	801	756	705	664
Sioux Falls	878	907	655	495	584	738	696	697	703
TENNESSEE									
Bristol	750	844	720	521	549	586	635	672	581
Chattanooga	827	909	798	613	589	695	651	686	618
Knoxville	808	909	781	588	613	683	682	674	610
Memphis	877	996	890	657	604	738	727	737	627
Nashville	871	923	811	602	597	655	695	685	620
TEXAS									
Abilene	899	1013	1098	1008	942	972	952	740	647
Amarillo	990	1072	1124	996	1018	990	957	844	692
Austin	839	982	883	785	703	728	745	644	516
Brownsville	767	959	841	732	703	729	658	567	489
Corpus Christi	878	1026	943	832	789	780	749	641	555
Dallas	911	999	961	886	824	845	848	740	593

Table 50   193

	Sept	Oct	Nov	Dec	Jan	Feb	Mar	Apr	May
El Paso	976	1164	1216	1077	1089	1136	1049	939	755
Fort Worth	911	999	961	886	824	845	848	740	593
Galveston	829	1032	978	802	831	741	712	643	560
Houston	767	940	883	936	713	671	680	590	524
Laredo	867	970	1006	898	886	904	854	672	515
Lubbock	956	1080	1135	1026	1032	997	950	833	696
Midland	976	1107	1186	1073	1074	995	972	829	686
Port Arthur	815	1010	929	753	745	716	732	655	574
San Angelo	883	1001	1077	1015	978	982	948	719	620
San Antonio	827	940	852	785	761	760	732	623	499
Victoria	818	1032	925	802	831	741	712	664	560
Waco	878	993	934	812	780	782	809	696	564
Wichita Falls	919	990	979	835	839	835	841	739	597
UTAH									
Salt Lake City	1101	1045	752	572	597	706	766	781	781
VERMONT									
Burlington	673	603	311	239	351	550	581	540	585
VIRGINIA									
Lynchburg	801	887	853	727	704	781	736	707	640
Norfolk	801	937	915	718	723	795	778	730	660
Richmond	807	922	851	673	683	768	771	740	681
Roanoke	763	858	720	562	563	572	624	695	591
WASHINGTON									
Olympia	724	538	347	248	297	474	580	595	610
Seattle	733	523	336	232	292	486	593	611	620
Spokane	918	743	315	194	252	535	702	773	749
Tacoma	716	523	324	226	285	468	571	589	620
Walla Walla	980	858	408	207	267	493	710	787	766
Yakima	1016	855	443	279	349	665	837	882	833
WEST VIRGINIA									
Charleston	743	778	557	431	460	517	554	551	549
Elkins	710	722	577	424	443	504	519	547	563
Huntington	769	806	587	444	474	544	566	587	590
Parkersburg	775	769	521	372	403	490	533	570	587
WISCONSIN									
Green Bay	766	729	485	399	461	652	665	643	640
La Crosse	813	840	531	404	520	649	636	685	658
Madison	780	779	516	408	488	621	610	597	623
Milwaukee	806	779	569	419	488	618	631	631	644
WYOMING									
Casper	910	876	705	610	720	849	844	732	655
Cheyenne	883	952	849	736	797	843	782	674	617

	Sept	*Oct*	*Nov*	*Dec*	*Jan*	*Feb*	*Mar*	*Apr*	*May*
Lander	936	932	768	666	733	901	881	743	698
Sheridan	884	841	642	518	586	791	750	700	673
ALBERTA									
Banff	585	504	289	117	182	386	421	461	450
Beaverlodge	612	517	246	129	178	390	552	624	650
Brooks	701	690	383	226	252	470	520	591	663
Calgary	660	663	402	265	300	503	533	555	559
Edmonton	669	627	333	198	237	435	599	637	669
Edson	566	561	250	169	224	445	524	541	563
Fairview	576	472	204	95	137	344	542	630	634
Lacombe	651	597	333	210	238	447	559	572	590
Lethbridge	764	715	428	297	334	553	589	601	642
Medicine Hat	672	661	381	254	271	505	506	577	631
Olds	665	652	356	219	249	469	534	627	555
Vauxhall	723	688	430	277	295	517	551	577	654
BRITISH COLUMBIA									
Campbell River	590	407	215	93	169	416	456	474	619
Dawson Creek	599	505	246	125	169	385	528	612	662
Fort St. John	603	495	232	111	157	373	530	618	670
Hope	531	393	167	114	151	318	356	346	414
Kamloops	690	469	210	96	159	372	552	504	570
Kitimat	450	239	120	75	133	274	462	461	530
Nanaimo	681	491	246	161	202	435	479	495	578
Nelson	686	435	210	79	115	369	510	462	610
New Westminster	654	509	279	161	177	435	493	495	601
Penticton	735	533	250	131	160	387	547	581	627
Port Alberni	681	491	246	161	196	424	467	495	578
Powell River	617	480	237	161	146	429	482	487	622
Prince George	543	392	184	92	138	345	482	534	618
Prince Rupert	343	209	141	62	125	258	326	349	380
Quesnel	559	410	194	101	154	360	502	543	632
Summerland	735	533	239	126	158	374	547	569	615
Terrace	439	239	117	75	133	274	462	461	530
Trail	697	435	152	42	84	322	467	421	532
Vancouver	654	509	279	161	182	435	493	495	630
Victoria	749	597	324	221	262	462	561	594	670
MANITOBA									
Brandon	656	648	299	239	298	486	524	562	563
Dauphin	620	624	304	242	317	516	560	614	627
Morden	681	675	325	267	312	551	516	568	578
Rivers	683	688	337	263	328	540	547	597	619
The Pas	604	527	151	133	178	410	530	555	606

Table 50   195

	Sept	Oct	Nov	Dec	Jan	Feb	Mar	Apr	May
Winnipeg	643	634	289	239	313	530	709	597	607
NEW BRUNSWICK									
Campbellton	580	499	307	266	320	507	511	506	528
Chatham	644	567	363	314	381	501	484	501	490
Fredericton	585	575	352	310	367	518	463	440	480
Moncton	585	575	374	315	373	518	451	463	495
Saint John	607	631	386	434	408	550	508	436	447
NEWFOUNDLAND									
Gander	505	445	250	198	240	365	344	316	354
St. John's	524	449	268	202	253	342	307	319	377
NOVA SCOTIA									
Annapolis Royal	620	575	332	192	241	376	479	448	479
Halifax	620	654	409	314	361	519	492	459	468
Kentville	607	575	324	212	285	428	423	436	479
Sydney	585	575	313	242	305	455	414	452	473
Truro	528	519	324	253	328	460	411	447	468
Yarmouth	590	600	378	230	286	413	463	489	525
ONTARIO									
Atikokan	628	418	230	200	296	591	634	545	501
Belleville	617	614	343	300	313	471	444	455	490
Brampton	668	656	406	311	367	496	468	500	577
Chatham	674	690	406	282	322	423	419	475	521
Cornwall	607	561	324	229	330	489	520	517	552
Guelph	590	586	330	259	302	433	432	478	544
Hamilton	599	591	332	283	371	528	472	503	594
Harrow	674	676	379	282	310	423	419	475	554
Kapuskasing	425	364	150	146	218	394	451	451	445
Kingston	671	628	394	300	367	509	468	489	599
Lindsay	603	572	317	208	287	458	468	478	501
London	626	612	323	236	266	419	417	473	548
Mount Forest	603	572	247	218	266	487	468	523	567
New Liskeard	462	391	179	155	223	418	489	472	407
North Bay	559	477	246	234	331	544	516	519	558
Ottawa	594	561	324	277	351	477	496	494	552
St. Catharines	652	577	345	196	232	335	422	468	517
Sarnia	652	707	311	229	313	463	447	529	558
Sault Ste. Marie	550	486	250	218	257	466	520	524	577
Thunder Bay	612	472	294	280	353	597	587	580	557
Toronto	698	652	365	294	338	478	488	520	535
Woodstock	663	640	336	247	266	393	428	484	516
PRINCE EDWARD ISLAND									
Charlottetown	639	547	301	210	302	455	451	429	463

	Sept	Oct	Nov	Dec	Jan	Feb	Mar	Apr	May
QUEBEC									
Amos	475	355	168	238	236	443	482	504	506
Berthierville	652	617	317	287	331	495	552	556	558
Grandes Bergeronnes	580	513	285	229	305	425	524	515	473
L'Assomption	625	589	328	265	331	495	504	521	558
Lennoxville	581	547	279	210	289	428	441	459	508
Maniwaki	550	499	261	223	309	506	508	558	587
Montreal	686	617	324	277	351	514	556	528	588
Normandin	505	432	233	215	281	438	497	497	501
Quebec	590	519	250	218	276	406	466	455	474
Rimouski	554	405	230	141	210	330	414	460	429
Ste. Agathe-des-Monts	585	540	236	269	322	520	492	553	592
St. Ambroise	492	364	203	208	289	438	493	553	501
Sept-Iles	564	486	325	295	317	539	516	609	525
Sherbrooke	607	575	311	229	299	428	460	482	519
Victoriaville	639	561	282	237	317	459	516	521	536
SASKATCHEWAN									
Estevan	784	769	426	311	377	551	622	603	682
Indian Head	737	661	347	232	298	521	567	585	642
Melfort	587	549	231	159	244	418	505	550	574
Moose Jaw	723	702	362	243	298	472	520	608	654
Outlook	634	624	322	209	289	463	545	579	675
Prince Albert	614	536	247	170	228	418	541	613	621
Regina	683	675	337	232	276	465	508	597	666
Saskatoon	732	691	324	212	260	494	618	644	684
Scott	651	571	281	155	220	421	480	596	648
Swift Current	683	675	381	250	287	472	506	585	642
Weyburn	777	621	411	258	321	487	579	608	701
Yorkton	660	464	258	232	275	473	549	602	639

Table 51   197

Table 51 AVERAGE DAILY B.T.U. OUTPUT INSIDE THE HOUSE FROM 90°-TILT DOUBLE-COVER COLLECTOR WITH HORIZONTAL REFLECTOR, PER SQUARE FOOT

	Sept	Oct	Nov	Dec	Jan	Feb	Mar	Apr	May
ALABAMA									
Birmingham	749	1007	899	653	652	745	669	621	475
Huntsville	748	1042	861	592	622	734	678	626	490
Mobile	683	1021	988	733	792	791	629	595	440
Montgomery	760	1027	1005	729	787	827	681	648	478
ARIZONA									
Flagstaff	900	1099	974	898	789	977	792	679	575
Phoenix	1010	1323	1333	1185	1153	1246	1015	867	670
Prescott	978	1227	1217	1031	1001	1104	904	789	640
Tucson	1006	1328	1422	1292	1240	1349	1018	867	638
Winslow	967	1242	1157	976	894	1075	915	818	648
Yuma	1040	1372	1472	1292	1318	1398	1103	905	666
ARKANSAS									
Fort Smith	806	1011	831	677	662	780	654	606	467
Little Rock	817	1134	861	663	620	810	703	626	505
Texarkana	851	1143	961	786	730	836	705	621	497
CALIFORNIA									
Bakersfield	1036	1302	1230	806	827	1110	947	807	640
Bishop	1009	1054	842	727	890	948	880	701	632
Burbank	942	1112	1190	1077	1102	1163	917	690	468
Eureka	620	709	600	502	545	654	687	568	471
Fresno	1091	1340	1100	663	672	991	939	851	720
Long Beach	942	1112	1190	1077	1102	1163	917	674	468
Los Angeles	908	1143	1254	1135	1102	1089	856	645	479
Oakland	948	1112	894	691	718	909	937	751	638
Red Bluff	1074	1196	932	664	671	932	916	812	705
Sacramento	1100	1271	958	596	600	889	920	809	709
San Diego	783	1033	1243	1132	1080	1076	805	636	412
San Francisco	829	1082	924	746	762	893	843	720	572
Santa Maria	898	1118	1119	988	962	1142	863	670	478
COLORADO									
Alamosa	868	821	779	616	712	793	683	520	479
Colorado Springs	861	997	865	865	868	961	869	621	503
Denver	857	1052	882	787	817	916	828	642	532
Grand Junction	920	1094	892	687	709	851	816	696	614
Pueblo	924	1120	996	896	931	1020	859	693	556
CONNECTICUT									
Bridgeport	757	812	689	588	596	691	670	558	554
Hartford	715	808	599	503	528	708	730	565	522
New Haven	769	915	703	635	620	813	747	609	545

	Sept	Oct	Nov	Dec	Jan	Feb	Mar	Apr	May
DELAWARE									
Wilmington	748	926	744	593	573	786	745	613	544
DISTRICT OF COLUMBIA									
Washington	741	923	767	597	611	747	714	607	528
FLORIDA									
Appalachicola	662	1028	1075	855	962	980	660	634	452
Daytona Beach	577	785	1034	1007	961	653	571	392	287
Jacksonville	629	822	1011	845	937	937	729	650	440
Key West	631	786	1151	1156	1191	1097	674	585	314
Miami Beach	633	772	1085	1131	1150	1075	664	568	305
Orlando	609	812	1034	1031	1057	1023	720	623	398
Pensacola	727	1064	1053	845	889	921	696	641	453
Tallahassee	652	973	1059	903	913	948	692	625	429
Tampa	659	883	1117	1022	1062	1042	710	618	388
GEORGIA									
Athens	783	1052	961	714	745	821	729	636	504
Atlanta	737	1007	930	671	701	806	681	626	490
Augusta	783	1062	952	780	796	847	717	674	487
Columbus	704	1062	967	750	796	863	717	645	501
Macon	760	989	1023	810	857	910	758	693	515
Rome	715	955	866	628	657	760	633	597	483
Savannah	628	882	997	851	881	874	709	657	471
IDAHO									
Boise	1069	961	561	395	443	651	807	757	653
Idaho Falls	967	810	478	300	349	524	646	620	572
Lewiston	973	862	383	189	302	517	691	705	620
Pocatello	761	721	493	354	390	523	631	585	510
ILLINOIS									
Cairo	880	1099	813	608	630	773	742	666	574
Chicago	815	896	596	448	505	658	690	586	577
Moline	840	911	558	470	516	605	647	565	513
Peoria	842	945	659	485	540	678	695	611	536
Rockford	827	896	571	452	505	580	685	576	558
Springfield	881	972	716	545	573	697	688	613	571
INDIANA									
Evansville	865	1011	719	544	579	593	520	499	475
Fort Wayne	781	856	530	428	453	596	657	568	554
Indianapolis	821	949	649	453	500	642	624	582	541
South Bend	802	867	482	382	424	528	659	565	577
IOWA									
Burlington	879	1034	762	575	587	746	760	643	563
Des Moines	802	940	672	517	614	721	710	617	568

Table 51   199

	Sept	Oct	Nov	Dec	Jan	Feb	Mar	Apr	May
Dubuque	747	779	543	431	526	660	659	607	549
Sioux City	840	940	672	517	614	760	735	649	577
Waterloo	849	907	581	470	556	693	731	614	581
KANSAS									
Concordia	869	1063	865	702	732	820	790	666	580
Dodge City	901	1132	968	845	885	798	643	432	482
Goodland	873	1050	879	782	843	947	844	661	566
Topeka	837	999	779	630	657	747	663	561	510
Wichita	865	1042	926	744	841	762	605	523	468
KENTUCKY									
Covington	813	908	629	482	530	649	645	582	536
Lexington	770	891	663	454	524	520	463	439	425
Louisville	806	966	705	505	565	569	492	455	454
LOUISIANA									
Alexandria	781	1064	1020	797	732	807	618	614	422
Baton Rouge	749	1021	988	765	792	794	629	559	416
Lake Charles	748	1042	1010	774	782	822	585	553	405
New Orleans	684	973	978	742	799	790	607	562	388
Shreveport	883	1136	1039	900	735	847	671	578	474
MAINE									
Caribou	556	555	307	238	285	460	508	456	432
Portland	767	828	610	545	589	747	752	578	553
MARYLAND									
Baltimore	751	918	784	631	653	796	690	581	494
Frederick	727	882	699	579	560	725	629	521	494
MASSACHUSETTS									
Boston	765	705	625	547	547	752	743	686	540
Nantucket	793	886	689	543	534	705	747	567	541
Pittsfield	629	705	473	324	459	639	654	520	494
Worcester	710	765	558	481	493	683	685	534	510
MICHIGAN									
Alpena	687	615	279	215	295	532	644	570	565
Detroit	765	793	456	317	396	541	609	544	531
Escanaba	697	665	362	328	392	622	681	575	534
Flint	729	705	373	282	358	626	627	540	531
Grand Rapids	746	732	385	307	313	504	615	572	563
Lansing	759	732	398	297	358	524	654	572	572
Marquette	638	522	264	216	293	481	591	539	508
Muskegon	759	702	373	297	324	498	654	583	540
Sault Ste. Marie	610	494	231	189	261	506	620	560	518
MINNESOTA									
Duluth	702	621	379	339	418	623	711	601	556

	Sept	Oct	Nov	Dec	Jan	Feb	Mar	Apr	May
International Falls	731	624	331	303	365	572	640	601	598
Minneapolis	793	755	465	374	490	639	691	596	589
Rochester	812	856	499	396	514	644	671	589	572
Saint Cloud	790	762	412	371	451	630	656	612	573
MISSISSIPPI									
Jackson	694	911	927	729	633	733	603	591	432
Meridian	749	969	927	714	757	780	647	610	451
Vicksburg	815	1027	965	684	710	780	652	601	451
MISSOURI									
Columbia	837	984	752	581	607	761	714	603	562
Kansas City	837	1014	806	643	711	804	752	639	553
St. Joseph	845	987	757	606	634	752	726	613	544
St. Louis	801	984	738	544	621	691	714	613	553
Springfield	833	978	811	635	639	767	690	615	510
MONTANA									
Billings	911	872	558	457	513	656	747	636	610
Glasgow	893	786	490	379	428	657	754	665	638
Great Falls	934	830	500	446	543	733	842	687	638
Havre	867	769	474	362	423	648	767	682	649
Helena	855	754	527	387	435	671	760	638	587
Kalispell	818	655	298	172	255	481	642	612	588
Miles City	895	865	571	459	485	734	767	697	752
Missoula	861	701	351	225	293	464	590	543	557
NEBRASKA									
Grand Island	867	989	762	620	674	806	777	640	581
Lincoln	818	975	782	620	669	800	773	643	563
Norfolk	852	940	698	539	625	773	761	638	586
North Platte	879	1034	801	654	740	854	802	640	572
Omaha	879	1063	723	575	652	819	752	671	607
Scottsbluff	878	963	761	667	722	887	824	649	568
NEVADA									
Elko	990	1025	754	586	658	813	777	626	584
Ely	946	1040	868	735	745	842	822	627	533
Las Vegas	1069	1291	1264	1038	1055	1172	990	833	664
Reno	1014	1086	895	678	720	899	902	764	656
Winnemucca	1026	1067	801	597	620	813	824	723	663
NEW HAMPSHIRE									
Concord	691	705	535	449	514	677	705	540	478
NEW JERSEY									
Atlantic City	784	820	814	647	649	800	758	624	553
Newark	781	922	730	588	609	780	754	632	567
Trenton	772	911	744	593	611	786	700	624	553

Table 51 201

	Sept	Oct	Nov	Dec	Jan	Feb	Mar	Apr	May
NEW MEXICO									
Albuquerque	932	1197	1128	962	960	1059	856	749	595
Raton	844	1029	953	798	894	996	825	627	502
Roswell	892	1067	1113	971	990	1048	836	725	522
Silver City	950	1168	1218	1083	1033	1164	914	772	556
NEW YORK									
Albany	746	790	485	402	481	652	679	561	534
Binghampton	589	632	378	284	361	502	520	460	458
Buffalo	772	746	385	297	358	524	628	520	540
New York City	781	922	730	588	625	780	734	632	567
Rochester	772	732	421	328	358	537	602	551	563
Schenectady	746	790	485	402	481	652	679	561	534
Syracuse	721	688	370	275	347	485	577	530	544
NORTH CAROLINA									
Asheville	714	958	876	660	676	780	666	601	482
Charlotte	760	1057	985	776	767	887	737	670	542
Greensboro	756	1006	895	728	708	821	722	593	531
Raleigh	732	961	910	720	713	851	722	643	523
Wilmington	726	1022	1047	860	865	882	764	700	526
Winston-Salem	732	796	851	706	666	762	685	623	508
NORTH DAKOTA									
Bismarck	841	810	513	413	462	667	694	607	577
Devils Lake	810	733	449	365	446	670	731	627	598
Fargo	814	783	411	390	418	623	694	617	616
Williston	893	786	490	389	436	669	744	676	669
OHIO									
Akron	757	812	438	282	346	488	579	537	554
Cincinnati	813	908	629	482	530	649	663	582	536
Cleveland	757	797	424	282	346	488	564	537	545
Columbus	796	889	595	424	439	601	624	571	549
Dayton	869	978	689	521	464	615	675	592	549
Mansfield	781	827	491	338	393	556	602	537	554
Sandusky	827	852	495	365	396	578	651	596	577
Toledo	777	808	482	328	367	554	621	555	549
Youngstown	745	768	413	282	334	529	564	516	536
OKLAHOMA									
Oklahoma City	852	1018	950	783	803	883	749	631	490
Tulsa	872	1036	984	809	680	792	661	543	438
OREGON									
Astoria	603	589	335	230	301	435	553	526	469
Eugene	845	611	364	225	285	459	568	557	480
Medford	827	735	495	293	340	553	642	565	503

	Sept	*Oct*	*Nov*	*Dec*	*Jan*	*Feb*	*Mar*	*Apr*	*May*
Pendleton	1015	862	453	214	380	504	691	694	710
Portland	754	603	347	241	301	463	589	557	520
Roseburg	875	615	371	210	292	450	562	560	534
Salem	812	609	357	218	300	456	553	564	490
PENNSYLVANIA									
Allentown	757	871	650	541	524	732	721	589	545
Erie	765	720	378	273	326	502	583	513	531
Harrisburg	748	880	635	521	547	711	700	603	544
Philadelphia	748	926	716	593	573	786	745	613	544
Pittsburgh	757	797	517	338	381	529	579	516	509
Reading	712	865	674	533	560	571	693	582	518
Scranton	696	783	517	417	429	624	618	516	500
Williamsport	671	738	464	349	405	569	618	506	482
RHODE ISLAND									
Providence	777	881	651	501	558	779	703	576	549
SOUTH CAROLINA									
Charleston	749	1003	1087	855	888	941	770	693	501
Columbia	726	1022	992	756	804	867	741	670	497
Florence	715	1007	976	742	789	852	729	660	490
Greenville	737	1011	955	719	767	871	752	706	535
Spartanburg	714	981	908	677	694	810	666	616	482
SOUTH DAKOTA									
Huron	872	853	605	458	584	734	754	648	638
Rapid City	890	926	708	553	633	785	783	640	572
Sioux Falls	864	912	629	483	565	724	721	633	605
TENNESSEE									
Bristol	692	888	711	502	534	613	616	595	477
Chattanooga	759	965	801	592	579	734	606	601	505
Knoxville	744	961	778	568	599	718	648	593	500
Memphis	806	1057	893	635	593	780	678	646	512
Nashville	802	976	807	581	583	689	661	603	508
TEXAS									
Abilene	816	1048	1123	991	943	1028	844	636	501
Amarillo	909	1138	1128	962	1001	1045	892	740	565
Austin	748	973	929	790	723	771	607	536	364
Brownsville	663	872	901	766	749	732	464	443	291
Corpus Christi	772	975	1001	855	826	805	570	518	362
Dallas	827	1033	983	870	825	894	752	636	460
El Paso	881	1186	1256	1067	1100	1202	904	798	569
Fort Worth	827	1033	983	870	825	894	752	636	460
Galveston	734	1001	1034	815	819	775	561	528	381
Houston	684	931	929	742	734	711	553	491	370

Table 51   203

	Sept	Oct	Nov	Dec	Jan	Feb	Mar	Apr	May
Laredo	762	922	1067	922	927	933	650	543	336
Lubbock	874	1131	1150	999	1024	1054	863	722	555
Midland	881	1128	1225	1063	1085	1053	837	704	517
Port Arthur	726	1000	978	758	766	758	596	544	405
San Angelo	792	1007	1123	1014	997	1039	795	605	453
San Antonio	737	931	896	790	782	806	596	518	352
Victoria	724	1001	978	815	819	775	561	545	381
Waco	793	1012	965	805	787	827	697	591	425
Wichita Falls	840	1037	992	814	832	882	764	641	475
UTAH									
Salt Lake City	1025	1078	723	552	572	718	786	702	652
VERMONT									
Burlington	674	601	299	234	340	532	603	491	508
VIRGINIA									
Lynchburg	739	933	842	701	684	817	714	625	526
Norfolk	739	986	904	692	703	832	754	646	542
Richmond	747	966	833	648	660	800	761	658	561
Roanoke	704	903	711	542	548	598	605	615	485
WASHINGTON									
Olympia	733	527	336	245	290	456	594	542	537
Seattle	745	508	326	231	287	467	603	555	550
Spokane	934	721	306	192	248	514	714	703	664
Tacoma	728	508	314	225	279	450	581	535	550
Walla Walla	987	848	394	204	260	476	732	716	670
Yakima	1028	838	429	276	341	641	857	803	733
WEST VIRGINIA									
Charleston	687	815	546	415	445	538	546	489	452
Elkins	657	754	560	408	427	522	521	488	465
Huntington	711	845	574	428	458	566	559	522	485
Parkersburg	717	802	506	358	388	508	535	509	484
WISCONSIN									
Green Bay	767	727	465	391	447	631	691	586	556
La Crosse	799	844	511	394	504	636	659	622	566
Madison	754	790	496	397	471	617	630	540	531
Milwaukee	779	790	547	407	471	613	651	571	549
WYOMING									
Casper	880	889	678	593	693	843	872	663	558
Cheyenne	823	982	817	710	763	858	802	606	515
Lander	905	945	738	647	706	894	910	673	595
Sheridan	886	839	617	508	569	766	779	638	584
ALBERTA									
Banff	596	481	284	118	181	371	419	434	405

	Sept	Oct	Nov	Dec	Jan	Feb	Mar	Apr	May
Beaverlodge	608	494	247	134	182	381	531	623	584
Brooks	715	658	376	228	251	452	518	557	596
Calgary	673	633	394	267	299	483	530	523	503
Edmonton	673	599	331	203	239	421	586	619	601
Edson	569	536	248	173	225	431	513	525	506
Fairview	568	451	206	100	141	337	517	637	569
La Combe	659	569	329	213	238	431	552	548	530
Lethbridge	784	682	419	298	331	529	591	557	577
Medicine Hat	690	630	373	255	269	483	508	535	567
Olds	673	622	351	223	249	452	528	513	499
Vauxhall	743	656	421	278	292	495	553	535	588
BRITISH COLUMBIA									
Campbell River	606	388	211	94	167	398	458	440	556
Dawson Creek	595	483	247	129	173	375	508	611	594
Fort St. John	594	474	234	117	162	365	506	625	602
Hope	542	378	163	113	149	305	360	315	369
Kamloops	704	447	206	97	159	357	550	475	512
Kitimat	450	228	120	77	134	266	449	454	476
Nanaimo	695	472	239	161	199	417	484	450	515
Nelson	704	414	205	80	114	353	512	429	548
New Westminster	667	490	272	161	174	417	498	450	536
Penticton	751	513	244	131	158	371	553	529	560
Port Alberni	695	472	239	161	194	406	472	450	515
Powell River	634	458	232	162	143	411	484	452	559
Prince George	542	374	184	95	140	335	468	526	555
Prince Rupert	343	200	141	63	127	250	317	344	342
Quesnel	563	391	193	103	155	348	491	528	568
Summerland	751	513	233	125	156	358	553	517	549
Terrace	439	228	117	77	134	266	449	454	476
Trail	712	418	148	42	82	309	472	382	475
Vancouver	667	490	272	161	179	417	498	450	562
Victoria	765	574	315	220	259	443	567	540	598
MANITOBA									
Brandon	674	618	292	240	296	465	526	521	506
Dauphin	632	595	298	244	316	495	558	579	564
Morden	695	649	316	266	308	528	522	517	515
Rivers	701	656	330	264	325	517	549	554	557
The Pas	604	503	151	137	180	398	515	547	544
Winnipeg	660	605	283	240	310	506	712	554	546
NEW BRUNSWICK									
Campbellton	590	485	298	264	314	487	520	460	469
Chatham	651	555	351	310	373	483	496	456	431

Table 51   205

	Sept	Oct	Nov	Dec	Jan	Feb	Mar	Apr	May
Fredericton	589	568	339	305	357	500	477	401	420
Moncton	589	568	360	310	363	500	465	421	433
Saint John	608	629	371	425	396	532	528	397	388
NEWFOUNDLAND									
Gander	516	429	244	198	237	350	347	287	316
St. John's	530	440	259	200	247	329	314	290	332
NOVA SCOTIA									
Annapolis Royal	621	573	319	188	234	364	498	408	415
Halifax	621	653	393	307	351	502	511	418	406
Kentville	608	573	311	208	276	414	440	397	415
Sydney	589	568	302	238	297	439	427	411	414
Truro	529	517	311	248	319	445	424	407	406
Yarmouth	580	603	363	224	277	405	480	444	452
ONTARIO									
Atikokan	641	403	224	200	293	567	640	496	447
Belleville	607	617	329	293	303	462	460	413	422
Brampton	658	660	390	303	355	487	485	454	497
Chatham	639	705	391	273	310	425	431	429	439
Cornwall	608	559	311	224	320	473	540	471	479
Guelph	580	589	317	253	292	424	448	434	469
Hamilton	579	600	319	275	358	524	487	455	506
Harrow	639	690	365	273	298	425	431	593	467
Kapuskasing	435	351	146	146	216	378	456	411	397
Kingston	660	631	378	293	355	499	485	444	513
Lindsay	593	575	305	203	277	449	485	434	431
London	605	621	311	230	257	416	430	428	467
Mount Forest	593	575	238	213	257	477	485	475	488
New Liskeard	469	380	173	154	219	402	498	429	361
North Bay	563	471	237	230	322	525	533	473	488
Ottawa	595	559	311	271	340	462	515	450	479
St. Catharines	630	585	332	191	224	332	436	424	441
Sarnia	630	717	298	222	302	460	461	479	476
Sault Ste. Marie	556	476	242	215	252	449	533	477	508
Thunder Bay	625	455	287	279	348	572	593	527	497
Toronto	686	655	351	287	327	468	506	472	461
Woodstock	641	649	323	240	257	390	443	438	439
PRINCE EDWARD ISLAND									
Charlottetown	643	540	290	206	294	439	465	390	405
QUEBEC									
Amos	483	344	163	236	232	426	490	458	449
Berthierville	656	609	305	283	322	478	570	506	488
Grandes Bergeronnes	590	498	277	227	299	408	533	468	420

	Sept	Oct	Nov	Dec	Jan	Feb	Mar	Apr	May
L'Assomption	630	582	317	260	322	478	520	475	488
Lennoxville	582	545	268	205	280	414	458	418	441
Maniwaki	556	489	253	221	302	487	520	508	517
Montreal	687	615	311	271	340	497	578	481	510
Normandin	516	416	227	215	277	420	502	452	447
Quebec	597	508	242	215	270	391	478	415	418
Rimouski	563	393	223	140	206	317	421	418	381
Ste. Agathe-des-Monts	589	533	228	265	313	501	508	503	518
St. Ambroise	502	351	198	207	285	420	498	503	447
Sept-Iles	576	468	316	295	313	517	522	554	469
Sherbrooke	608	573	299	224	290	414	477	439	450
Victoriaville	643	554	271	233	309	443	533	475	469
SASKATCHEWAN									
Estevan	801	740	415	311	373	528	629	548	608
Indian Head	756	630	340	233	296	498	569	543	577
Melfort	590	524	229	162	246	405	494	534	516
Moose Jaw	743	669	354	244	295	451	522	565	588
Outlook	646	595	316	211	288	445	543	545	607
Prince Albert	618	512	245	174	229	405	529	596	558
Regina	701	643	330	233	273	444	510	554	598
Saskatoon	742	659	320	215	260	476	610	616	615
Scott	659	545	277	157	220	404	474	570	582
Swift Current	701	643	373	251	284	451	508	543	577
Weyburn	798	592	402	259	319	466	581	565	630
Yorkton	673	443	253	234	274	454	546	568	574

Notes to Tables 50 and 51: *(1) Calculation procedure described in Chapter 13 used, taking values from Tables 15, 46, 42, 44, and 52. (2) One square foot of collector area, after subtracting for glazing support (1 ft² net). (3) R-30-insulated collector and storage.*

Appendix B

Table 52 OVER-ALL SYSTEM EFFICIENCY OF TWO-COVER R-30-INSULATED COLLECTOR WITH REFLECTOR AND R-30-INSULATED STORAGE UNDER 155 B.T.U./FT^2·HR INSOLATION TRANSMITTED THROUGH COVERS

	Sept	Oct	Nov	Dec	Jan	Feb	Mar	Apr	May
ALABAMA									
Birmingham	87	87	83	80	80	80	83	87	87
Huntsville	87	87	80	77	77	80	83	87	87
Mobile	87	87	85	83	83	83	85	87	87
Montgomery	87	87	83	80	80	83	83	87	87
ARIZONA									
Flagstaff	85	80	75	70	70	73	73	77	83
Phoenix	87	87	85	83	80	83	85	87	87
Prescott	87	85	80	75	75	77	77	83	85
Tucson	87	87	85	83	80	83	85	87	87
Winslow	87	85	77	73	70	75	77	85	87
Yuma	87	87	87	83	83	85	87	87	87
ARKANSAS									
Fort Smith	87	87	80	77	75	80	80	87	87
Little Rock	87	87	80	77	75	80	83	87	87
Texarkana	87	87	83	80	77	80	83	87	87
CALIFORNIA									
Bakersfield	87	87	85	80	80	83	85	87	87
Bishop	87	85	80	75	75	77	80	83	85
Burbank	87	87	85	83	83	85	85	87	87
Eureka	85	83	83	80	80	80	80	83	83
Fresno	87	87	83	80	80	83	83	85	87
Long Beach	87	87	85	83	83	85	85	85	87
Los Angeles	87	87	85	85	83	83	85	85	85
Oakland	87	87	83	80	80	83	83	85	85
Red Bluff	87	87	83	80	77	83	83	85	87
Sacramento	87	87	83	80	77	83	83	85	87
San Diego	87	87	87	85	83	85	85	87	87
San Francisco	87	87	83	80	80	83	83	85	85
Santa Maria	87	87	85	83	80	83	83	85	85
COLORADO									
Alamosa	85	77	70	66	66	70	73	77	83
Colorado Springs	87	80	75	73	70	73	75	80	83
Denver	87	83	75	73	70	73	75	80	85
Grand Junction	87	83	75	70	69	73	77	83	87
Pueblo	87	83	75	73	70	75	75	83	85

	Sept	Oct	Nov	Dec	Jan	Feb	Mar	Apr	May
CONNECTICUT									
Bridgeport	87	83	77	73	70	73	75	80	85
Hartford	87	83	77	70	69	70	75	80	85
New Haven	87	83	77	73	70	73	75	80	85
DELAWARE									
Wilmington	87	85	80	73	73	75	77	83	87
DISTRICT OF COLUMBIA									
Washington	87	85	80	75	75	75	77	85	87
FLORIDA									
Appalachicola	87	87	85	83	83	85	85	87	87
Daytona Beach	87	87	87	85	85	85	87	87	87
Jacksonville	87	87	87	83	83	85	85	87	87
Key West	87	87	87	87	87	87	87	87	87
Miami Beach	87	87	87	87	87	87	87	87	87
Orlando	87	87	87	85	85	85	87	87	87
Pensacola	87	87	85	83	83	85	85	87	87
Tallahassee	87	87	85	83	83	85	85	87	87
Tampa	87	87	87	85	85	87	87	87	87
GEORGIA									
Athens	87	87	83	77	77	80	83	85	87
Atlanta	87	87	83	77	77	80	83	85	87
Augusta	87	87	83	80	80	83	83	87	87
Columbus	87	87	83	80	80	83	83	87	87
Macon	87	87	85	80	80	83	85	87	87
Rome	87	85	80	77	77	80	80	85	87
Savannah	87	87	85	83	83	83	85	87	87
IDAHO									
Boise	87	83	75	73	70	75	77	83	85
Idaho Falls	85	77	70	66	65	69	70	77	83
Lewiston	87	83	75	73	70	75	77	83	85
Pocatello	66	62	62	65	65	61	62	69	70
ILLINOIS									
Cairo	87	85	80	75	75	77	80	85	87
Chicago	87	83	75	70	69	73	75	80	85
Moline	87	83	75	70	69	70	73	80	85
Peoria	87	83	75	70	69	73	75	83	85
Rockford	87	83	75	69	69	70	73	80	85
Springfield	87	85	77	73	70	73	75	83	87
INDIANA									
Evansville	87	85	77	75	73	75	77	85	87
Fort Wayne	87	83	75	70	70	73	75	80	85
Indianapolis	87	83	77	70	70	73	75	83	85

Table 52 209

	Sept	Oct	Nov	Dec	Jan	Feb	Mar	Apr	May
South Bend	87	83	75	70	69	70	73	80	85
IOWA									
Burlington	87	83	75	70	69	73	75	83	85
Des Moines	87	83	75	69	66	70	73	80	85
Dubuque	85	80	73	69	66	69	73	80	85
Sioux City	87	83	75	69	66	70	73	80	85
Waterloo	87	83	73	69	66	69	73	80	85
KANSAS									
Concordia	87	85	77	73	70	73	75	83	87
Dodge City	87	85	77	73	70	75	77	66	87
Goodland	87	83	75	70	70	73	75	80	85
Topeka	87	85	77	73	70	75	77	83	87
Wichita	87	85	77	73	73	75	77	85	87
KENTUCKY									
Covington	87	85	77	73	73	75	75	83	87
Lexington	87	85	77	73	73	75	77	83	87
Louisville	87	85	77	75	73	75	77	83	87
LOUISIANA									
Alexandria	87	87	85	83	80	83	85	87	87
Baton Rouge	87	87	85	83	83	85	85	87	87
Lake Charles	87	87	85	83	83	85	85	87	87
New Orleans	87	87	85	83	83	85	85	87	87
Shreveport	87	87	85	80	83	80	83	87	87
MAINE									
Caribou	83	77	73	65	62	66	69	75	80
Portland	85	80	75	69	69	70	73	77	83
MARYLAND									
Baltimore	87	85	80	73	73	75	77	83	87
Frederick	87	83	77	73	73	75	77	83	87
MASSACHUSETTS									
Boston	87	83	77	73	70	73	75	80	85
Nantucket	87	83	80	75	73	73	75	77	83
Pittsfield	85	80	75	69	69	70	73	77	83
Worcester	85	80	75	70	69	70	73	77	83
MICHIGAN									
Alpena	85	80	73	69	66	69	69	75	83
Detroit	87	83	77	70	70	70	73	80	85
Escanaba	85	80	73	69	66	69	69	75	80
Flint	85	80	75	69	69	70	70	77	83
Grand Rapids	87	83	75	70	69	69	73	80	85
Lansing	87	83	75	70	69	70	73	80	85
Marquette	85	80	73	69	66	69	70	75	80

	Sept	Oct	Nov	Dec	Jan	Feb	Mar	Apr	May
Muskegon	87	83	75	70	69	70	73	80	83
Sault Ste. Marie	85	80	73	66	65	66	69	75	80
MINNESOTA									
Duluth	83	77	70	65	62	65	66	75	80
International Falls	83	77	69	62	61	65	66	75	83
Minneapolis	85	80	73	66	65	66	70	77	85
Rochester	85	80	73	66	65	66	70	77	83
Saint Cloud	85	80	70	65	62	66	69	77	83
MISSISSIPPI									
Jackson	87	87	83	80	80	83	83	87	87
Meridian	87	87	83	80	80	83	83	87	87
Vicksburg	87	87	85	80	80	83	85	87	87
MISSOURI									
Columbia	87	85	77	73	70	75	77	83	87
Kansas City	87	85	77	73	73	75	77	85	87
St. Joseph	87	85	77	73	70	73	75	83	87
St. Louis	87	85	77	73	73	75	77	83	87
Springfield	87	85	77	75	73	75	77	85	87
MONTANA									
Billings	85	80	75	70	69	70	73	80	85
Glasgow	85	77	70	66	62	66	70	77	83
Great Falls	85	80	73	70	69	70	70	77	83
Havre	83	80	70	66	65	66	70	77	83
Helena	85	77	73	69	66	70	73	77	83
Kalispell	83	77	73	69	66	70	73	77	83
Miles City	85	80	73	69	65	69	70	80	85
Missoula	83	77	73	69	66	70	73	77	83
NEBRASKA									
Grand Island	87	83	75	70	66	70	73	80	85
Lincoln	87	83	77	70	69	73	75	83	85
Norfolk	87	83	75	69	66	70	73	80	85
North Platte	87	83	75	70	69	73	73	80	85
Omaha	87	83	75	70	66	70	73	80	85
Scottsbluff	87	80	75	70	69	73	73	80	85
NEVADA									
Elko	85	80	73	70	69	73	73	77	83
Ely	85	80	73	70	69	70	73	77	80
Las Vegas	87	87	83	77	77	80	83	87	87
Reno	85	80	75	73	70	75	77	80	83
Winnemucca	85	80	75	70	70	73	75	80	83
NEW HAMPSHIRE									
Concord	85	80	75	69	66	70	73	77	85

Table 52   211

	Sept	Oct	Nov	Dec	Jan	Feb	Mar	Apr	May
NEW JERSEY									
Atlantic City	87	85	80	75	73	75	77	83	87
Newark	87	85	80	73	73	75	77	83	87
Trenton	87	85	80	73	73	75	75	83	87
NEW MEXICO									
Albuquerque	87	85	77	75	73	77	80	85	87
Raton	87	83	75	73	70	73	75	80	83
Roswell	87	85	80	75	75	77	80	85	87
Silver City	87	85	80	77	75	77	80	85	87
NEW YORK									
Albany	87	83	75	70	69	70	73	80	85
Binghampton	87	83	77	70	70	70	73	80	85
Buffalo	87	83	75	70	69	70	73	77	83
New York City	87	85	80	73	75	75	75	83	87
Rochester	87	83	77	70	69	70	73	80	85
Schenectady	87	83	75	70	69	70	73	80	85
Syracuse	87	83	77	70	69	70	73	80	85
NORTH CAROLINA									
Asheville	87	85	80	75	75	77	80	85	87
Charlotte	87	87	83	77	77	80	80	85	87
Greensboro	87	85	80	75	75	77	80	85	87
Raleigh	87	85	80	77	77	77	80	85	87
Wilmington	87	87	85	80	80	80	83	87	87
Winston-Salem	87	85	80	77	75	77	80	85	87
NORTH DAKOTA									
Bismarck	85	80	70	66	62	66	69	77	83
Devils Lake	85	77	70	65	61	65	69	75	83
Fargo	85	80	70	65	62	65	69	77	83
Williston	85	77	70	65	62	66	69	77	83
OHIO									
Akron	87	83	77	70	70	73	75	80	85
Cincinnati	87	85	77	73	73	75	77	83	87
Cleveland	87	83	77	70	70	73	73	80	85
Columbus	87	83	77	73	70	73	75	83	85
Dayton	87	83	77	73	70	73	75	83	85
Mansfield	87	83	75	70	70	73	73	80	85
Sandusky	87	83	77	73	70	73	75	80	85
Toledo	87	83	75	70	69	70	73	80	85
Youngstown	87	83	75	70	70	73	73	80	85
OKLAHOMA									
Oklahoma City	87	85	80	75	75	77	80	85	87
Tulsa	87	85	80	75	75	77	80	85	87

	Sept	Oct	Nov	Dec	Jan	Feb	Mar	Apr	May
OREGON									
Astoria	85	83	80	77	75	77	77	80	83
Eugene	87	83	80	77	75	77	80	83	85
Medford	87	83	77	75	73	77	77	83	85
Pendleton	87	83	77	75	73	75	77	83	85
Portland	87	83	80	77	75	77	80	83	85
Roseburg	87	83	80	77	75	77	80	83	85
Salem	87	83	80	77	75	77	77	83	85
PENNSYLVANIA									
Allentown	87	83	77	73	70	73	75	80	85
Erie	87	83	77	70	70	70	73	80	85
Harrisburg	87	85	77	73	73	73	75	83	87
Philadelphia	87	85	77	73	73	75	77	83	87
Pittsburgh	87	83	77	70	70	73	75	80	85
Reading	87	85	80	73	73	61	77	83	87
Scranton	87	83	75	70	70	73	75	80	85
Williamsport	87	83	77	70	70	73	75	80	85
RHODE ISLAND									
Providence	87	83	77	73	70	73	75	80	85
SOUTH CAROLINA									
Charleston	87	87	85	80	80	83	85	87	87
Columbia	87	87	83	80	80	80	83	87	87
Florence	87	87	83	80	80	80	83	87	87
Greenville	87	87	83	77	77	80	83	87	87
Spartanburg	87	87	83	77	77	80	80	87	87
SOUTH DAKOTA									
Huron	85	80	73	66	69	66	70	77	85
Rapid City	85	80	75	70	69	70	70	77	83
Sioux Falls	85	80	73	66	65	69	70	80	85
TENNESSEE									
Bristol	87	85	80	75	75	77	80	85	87
Chattanooga	87	87	80	77	77	80	80	85	87
Knoxville	87	85	80	77	77	77	80	85	87
Memphis	87	87	83	77	77	80	80	87	87
Nashville	87	85	80	77	75	77	80	85	87
TEXAS									
Abilene	87	87	83	80	77	80	83	87	87
Amarillo	87	85	80	75	75	77	80	85	87
Austin	87	87	85	83	80	83	85	87	87
Brownsville	87	87	87	87	85	87	87	87	87
Corpus Christi	87	87	87	85	85	85	87	87	87
Dallas	87	87	83	80	77	83	83	87	87

Table 52   213

	Sept	Oct	Nov	Dec	Jan	Feb	Mar	Apr	May
El Paso	87	87	83	77	77	83	83	87	87
Fort Worth	87	87	83	80	77	83	83	87	87
Galveston	87	87	87	85	83	85	85	87	87
Houston	87	87	85	83	83	85	85	87	87
Laredo	87	87	87	85	85	87	87	87	87
Lubbock	87	85	80	77	75	77	80	85	87
Midland	87	87	83	80	77	80	83	87	87
Port Arthur	87	87	85	83	83	85	85	87	87
San Angelo	87	87	83	80	80	83	85	87	87
San Antonio	87	87	85	83	83	85	85	87	87
Victoria	87	87	85	83	83	85	85	87	87
Waco	87	87	85	80	80	83	85	87	87
Wichita Falls	87	87	83	77	77	80	83	87	87
UTAH									
Salt Lake City	87	83	75	70	70	73	75	80	85
VERMONT									
Burlington	85	80	75	69	65	69	70	77	83
VIRGINIA									
Lynchburg	87	85	80	75	75	77	77	85	87
Norfolk	87	87	83	77	77	77	80	85	87
Richmond	87	85	80	75	75	77	80	85	87
Roanoke	87	85	80	75	75	77	77	85	87
WASHINGTON									
Olympia	85	83	77	75	75	77	77	80	83
Seattle	87	83	80	77	77	80	80	83	85
Spokane	85	80	75	70	69	73	75	80	85
Tacoma	85	83	77	75	75	77	77	80	85
Walla Walla	87	83	77	75	73	77	80	83	85
Yakima	87	80	75	73	70	75	77	83	85
WEST VIRGINIA									
Charleston	87	85	80	75	75	75	77	83	87
Elkins	87	83	77	73	73	75	75	83	85
Huntington	87	85	80	75	75	75	77	85	87
Parkersburg	87	85	77	73	73	75	77	83	87
WISCONSIN									
Green Bay	85	80	73	69	66	69	70	77	83
La Crosse	85	83	73	69	65	69	70	80	85
Madison	85	80	73	69	66	69	69	77	83
Milwaukee	85	80	75	69	66	70	70	77	83
WYOMING									
Casper	85	80	73	70	69	70	73	77	83
Cheyenne	85	80	73	70	69	70	73	77	83

	Sept	Oct	Nov	Dec	Jan	Feb	Mar	Apr	May
Lander	85	80	73	69	66	70	73	77	83
Sheridan	85	80	73	69	66	70	70	77	83
ALBERTA									
Banff	80	75	70	65	65	69	69	75	77
Beaverlodge	80	75	69	62	61	66	69	75	80
Brooks	83	77	70	65	61	66	69	75	83
Calgary	83	77	70	66	65	69	69	75	80
Edmonton	83	77	69	65	61	66	69	75	83
Edson	80	75	69	65	61	66	69	75	80
Fairview	80	75	69	62	58	65	66	75	80
Lacombe	83	75	69	65	61	66	69	75	80
Lethbridge	83	77	73	69	65	69	70	77	83
Medicine Hat	85	77	70	66	62	69	69	77	83
Olds	83	77	70	66	62	66	69	75	80
Vauxhall	83	77	70	66	62	69	70	77	83
BRITISH COLUMBIA									
Campbell River	85	80	77	75	73	77	75	80	83
Dawson Creek	80	75	69	62	58	65	66	75	80
Fort St. John	80	75	69	62	58	65	66	75	80
Hope	85	83	77	73	70	75	77	80	85
Kamloops	85	80	75	70	69	73	75	80	85
Kitimat	85	77	75	70	69	73	75	77	83
Nanaimo	85	80	77	75	75	77	77	80	83
Nelson	85	80	75	70	69	73	75	80	85
New Westminster	85	83	77	75	73	77	77	80	83
Penticton	85	80	75	73	70	73	75	80	85
Port Alberni	85	80	77	75	73	75	75	80	83
Powell River	85	83	77	75	75	77	77	80	85
Prince George	80	75	70	66	62	69	70	75	80
Prince Rupert	83	80	77	75	73	75	75	77	80
Quesnel	83	77	70	66	65	70	73	77	83
Summerland	85	80	75	70	69	73	75	80	85
Terrace	83	77	73	70	69	73	75	77	83
Trail	87	80	75	70	69	73	75	80	85
Vancouver	85	83	77	75	75	77	77	80	87
Victoria	85	83	77	77	75	77	77	80	83
MANITOBA									
Brandon	83	77	69	62	58	62	65	75	80
Dauphin	83	77	69	62	58	62	65	75	80
Morden	85	77	70	62	58	65	66	75	83
Rivers	83	77	69	61	58	62	65	75	80
The Pas	83	75	66	58	55	61	65	73	80

Table 52   215

	Sept	Oct	Nov	Dec	Jan	Feb	Mar	Apr	May
Winnipeg	83	77	69	62	58	62	66	75	83
NEW BRUNSWICK									
Campbellton	83	77	73	66	65	66	69	75	80
Chatham	85	77	73	66	65	66	69	75	80
Fredericton	85	80	73	66	65	69	70	75	83
Moncton	85	80	75	69	66	69	70	75	80
Saint John	85	80	75	75	66	69	70	75	80
NEWFOUNDLAND									
Gander	83	77	75	69	66	69	69	73	77
St. John's	83	80	75	73	69	70	70	75	77
NOVA SCOTIA									
Annapolis Royal	85	80	77	70	69	70	73	77	80
Halifax	85	83	77	70	69	70	73	77	80
Kentville	85	80	75	70	69	70	70	75	80
Sydney	85	80	75	70	69	69	70	75	80
Truro	85	80	75	70	69	69	70	75	80
Yarmouth	85	80	77	73	70	70	73	77	80
ONTARIO									
Atikokan	80	77	69	62	58	62	66	75	80
Belleville	87	80	75	69	66	69	70	77	83
Brampton	85	80	75	69	66	69	70	77	83
Chatham	87	83	77	70	69	70	73	80	85
Cornwall	85	80	75	66	65	66	70	77	83
Guelph	85	80	75	69	66	69	70	77	83
Hamilton	87	83	77	70	69	70	73	80	85
Harrow	87	83	77	70	69	70	73	80	85
Kapuskasing	83	77	69	61	58	62	65	73	80
Kingston	87	80	75	69	66	69	70	77	83
Lindsay	85	80	75	66	65	69	70	77	83
London	87	80	75	69	66	69	70	77	83
Mount Forest	85	80	73	66	65	66	70	77	80
New Liskeard	83	77	70	65	61	65	66	75	80
North Bay	85	80	73	65	62	66	69	75	83
Ottawa	85	80	75	66	65	66	70	77	83
St. Catharines	87	83	77	70	69	70	73	80	85
Sarnia	87	83	75	70	69	70	73	77	83
Sault Ste. Marie	83	77	73	66	65	66	69	75	80
Thunder Bay	83	77	70	65	61	65	66	75	80
Toronto	87	83	77	70	69	70	73	80	85
Woodstock	85	80	75	69	66	69	70	77	83
PRINCE EDWARD ISLAND									
Charlottetown	85	80	75	69	66	69	70	75	80

	Sept	Oct	Nov	Dec	Jan	Feb	Mar	Apr	May
QUEBEC									
Amos	83	75	70	70	58	62	65	73	80
Berthierville	85	80	73	65	62	66	69	77	83
Grand Bergeronnes	83	77	73	65	62	66	69	73	80
L'Assomption	85	80	73	66	62	66	69	77	83
Lennoxville	85	80	73	66	65	66	69	77	83
Maniwaki	83	77	73	65	62	65	69	75	83
Montreal	85	80	75	66	65	66	70	77	85
Normandin	83	77	70	62	58	62	65	73	80
Quebec	85	80	73	66	65	66	70	75	83
Rimouski	83	77	73	66	65	66	69	75	80
Ste. Agathe-des-Monts	83	77	70	65	62	66	69	75	83
St. Ambroise	83	77	70	62	58	62	66	73	80
Sept-Iles	80	75	70	65	62	65	66	73	77
Sherbrooke	85	80	75	66	65	66	70	77	83
Victoriaville	85	80	73	66	65	66	69	77	83
SASKATCHEWAN									
Estevan	83	77	70	62	61	65	66	75	83
Indian Head	83	77	69	62	58	65	66	75	83
Melfort	83	75	66	61	58	62	65	73	80
Moose Jaw	83	77	70	65	61	66	66	75	83
Outlook	83	77	69	62	61	65	66	75	83
Prince Albert	83	75	66	61	57	62	65	75	80
Regina	83	77	69	62	58	65	66	75	83
Saskatoon	83	77	69	62	58	65	66	75	83
Scott	83	75	69	61	58	62	65	75	80
Swift Current	83	77	70	65	62	66	69	75	83
Weyburn	83	77	70	65	61	65	66	75	83
Yorkton	83	77	69	61	58	62	65	75	80

Note to Table 52: *Derived from heuristic values at various outside temperatures. It may be of interest to the serious student of applications to compare these values with the ones obtained by multiplying the efficiencies appearing in Table 23 by those in Table 28, as against outside temperatures.*

Appendix C

Table 53 AVERAGE DEGREE DAYS PER DAY, TAKEN OVER THE 273-DAY HEATING SEASON

	Average Degree-Days Per Day		Average Degree-Days Per Day		Average Degree-Days Per Day
ALABAMA		Pueblo	19.97	Moline	23.30
Birmingham	9.34	**CONNECTICUT**		Peoria	21.93
Huntsville	11.25	Bridgeport	20.48	Rockford	24.74
Mobile	5.71	Hartford	22.50	Springfield	19.82
Montgomery	8.39	New Haven	21.39	**INDIANA**	
ARIZONA		**DELAWARE**		Evansville	16.25
Flagstaff	25.12	Wilmington	18.04	Fort Wayne	22.55
Phoenix	6.47	**DISTRICT OF COLUMBIA**		Indianapolis	20.73
Prescott	15.92	Washington	15.47	South Bend	23.34
Tucson	6.59	**FLORIDA**		**IOWA**	
Winslow	17.52	Appalachicola	4.79	Burlington	22.27
Yuma	4.46	Daytona Beach	3.22	Des Moines	24.76
ARKANSAS		Jacksonville	4.54	Dubuque	26.58
Fort Smith	12.06	Key West	0.40	Sioux City	25.29
Little Rock	11.77	Miami Beach	0.52	Waterloo	26.50
Texarkana	9.29	Orlando	2.81	**KANSAS**	
CALIFORNIA		Pensacola	5.36	Concordia	20.00
Bakersfield	7.77	Tallahassee	5.44	Dodge City	18.23
Bishop	15.35	Tampa	2.50	Goodland	22.32
Burbank	5.96	**GEORGIA**		Topeka	18.94
Eureka	14.03	Athens	10.73	Wichita	16.90
Fresno	9.13	Atlanta	10.93	**KENTUCKY**	
Long Beach	6.20	Augusta	8.78	Covington	19.20
Los Angeles	7.07	Columbus	8.73	Lexington	17.15
Oakland	9.81	Macon	7.82	Louisville	17.04
Red Bluff	9.21	Rome	12.18	**LOUISIANA**	
Sacramento	10.14	Savannah	6.66	Alexandria	7.04
San Diego	5.12	**IDAHO**		Baton Rouge	5.71
San Francisco	10.00	Boise	20.98	Lake Charles	5.34
Santa Maria	9.56	Idaho Falls	31.18	New Orleans	5.07
COLORADO		Lewiston	19.97	Shreveport	8.00
Alamosa	30.03	Pocatello	25.25	**MAINE**	
Colorado Springs	23.10	**ILLINOIS**		Caribou	34.40
Denver	22.72	Cairo	14.00	Portland	26.87
Grand Junction	20.59	Chicago	22.37	**MARYLAND**	

	Average Degree-Days Per Day		*Average Degree-Days Per Day*		*Average Degree-Days Per Day*
Baltimore	17.05	Miles City	27.88	Winston-Salem	13.17
Frederick	18.59	Missoula	28.56	NORTH DAKOTA	
MASSACHUSETTS		NEBRASKA		Bismark	31.77
Boston	20.47	Grand Island	23.73	Devils Lake	35.42
Nantucket	20.98	Lincoln	21.35	Fargo	33.19
Pittsfield	27.07	Norfolk	25.36	Williston	33.07
Worcester	25.10	North Platte	24.25	OHIO	
MICHIGAN		Omaha	24.02	Akron	21.94
Alpena	29.95	Scottsbluff	40.19	Cincinnati	17.57
Detroit	22.67	NEVADA		Cleveland	22.90
Escanaba	29.95	Elko	26.37	Columbus	20.61
Flint	26.49	Ely	27.34	Dayton	20.46
Grand Rapids	24.84	Las Vegas	9.92	Mansfield	23.12
Lansing	24.95	Reno	22.03	Sandusky	21.08
Marquette	29.58	Winnemucca	24.08	Toledo	23.51
Muskegon	24.10	NEW HAMPSHIRE		Youngstown	23.19
Sault Ste. Marie	31.67	Concord	26.56	OKLAHOMA	
MINNESOTA		NEW JERSEY		Oklahoma City	13.64
Duluth	35.25	Atlantic City	17.57	Tulsa	14.14
International Falls	37.54	Newark	17.80	OREGON	
Minneapolis	30.21	Trenton	18.20	Astoria	17.14
Rochester	29.83	NEW MEXICO		Eugene	16.57
Saint Cloud	31.86	Albuquerque	15.93	Medford	18.06
MISSISSIPPI		Raton	22.45	Pendleton	18.55
Jackson	8.20	Roswell	13.89	Portland	16.40
Meridian	8.38	Silver City	13.57	Roseburg	15.86
Vicksburg	7.48	NEW YORK		Salem	16.64
MISSOURI		Albany	24.95	PENNSYLVANIA	
Columbia	18.44	Binghampton	23.36	Allentown	21.19
Kansas City	17.26	Buffalo	25.38	Erie	23.32
St. Joseph	20.01	New York City	17.81	Harrisburg	19.19
St. Louis	17.89	Rochester	24.40	Philadelphia	18.64
Springfield	16.68	Schenectady	24.17	Pittsburgh	21.75
MONTANA		Syracuse	24.46	Reading	18.11
Billings	25.37	NORTH CAROLINA		Scranton	22.72
Glasgow	32.12	Asheville	14.81	Williamsport	21.62
Great Falls	27.41	Charlotte	11.69	RHODE ISLAND	
Havre	30.98	Greensboro	13.94	Providence	21.56
Helena	28.73	Raleigh	12.43	SOUTH CAROLINA	
Kalispell	28.70	Wilmington	8.60	Charleston	7.45

Table 53   219

	Average Degree-Days Per Day		Average Degree-Days Per Day		Average Degree-Days Per Day
Columbia	9.10	Norfolk	12.53	Campbell River	21.30
Florence	8.74	Richmond	14.16	Dawson Creek	38.34
Greenville	10.56	Roanoke	15.20	Fort St. John	38.72
Spartanburg	11.26	WASHINGTON		Hope	20.27
SOUTH DAKOTA		Olympia	18.02	Kamloops	23.81
Huron	29.73	Seattle	15.42	Kitimat	25.80
Rapid City	26.32	Spokane	23.76	Nanaimo	19.42
Sioux Falls	28.27	Tacoma	17.83	Nelson	23.82
TENNESSEE		Walla Walla	17.44	New Westminster	18.90
Bristol	15.18	Yakima	21.47	Penticton	23.12
Chattanooga	11.92	WEST VIRGINIA		Port Alberni	21.10
Knoxville	12.80	Charleston	16.36	Powell River	18.23
Memphis	11.84	Elkins	20.49	Prince George	33.50
Nashville	13.11	Huntington	16.24	Prince Rupert	22.41
TEXAS		Parkersburg	17.39	Quesnel	30.20
Abilene	9.61	WISCONSIN		Summerland	23.22
Amarillo	14.60	Green Bay	28.76	Terrace	26.40
Austin	6.27	La Crosse	27.43	Trail	23.29
Brownsville	2.20	Madison	28.19	Vancouver	18.04
Corpus Christi	3.35	Milwaukee	27.14	Victoria	17.85
Dallas	8.66	WYOMING		MANITOBA	
El Paso	9.89	Casper	26.59	Brandon	39.27
Fort Worth	8.81	Cheyenne	26.10	Dauphin	39.41
Galveston	4.52	Lander	28.18	Morden	36.53
Houston	5.11	Sheridan	27.39	Rivers	39.26
Laredo	2.92	CANADA		The Pas	43.52
Lubbock	13.11	ALBERTA		Winnipeg	38.32
Midland	9.49	Banff	35.21	NEW BRUNSWICK	
Port Arthur	5.30	Beaverlodge	37.15	Campbellton	33.11
San Angelo	8.26	Brooks	34.38	Chatham	31.62
San Antonio	5.67	Calgary	33.71	Fredericton	30.21
Victoria	4.30	Edmonton	35.78	Moncton	30.33
Waco	7.44	Edson	36.93	Saint John	26.75
Wichita Falls	10.37	Fairview	38.99	NEWFOUNDLAND	
UTAH		Lacombe	36.60	Gander	31.38
Salt Lake City	21.86	Lethbridge	30.93	St. John's	28.33
VERMONT		Medicine Hat	31.82	NOVA SCOTIA	
Burlington	29.62	Olds	34.88	Annapolis Royal	26.12
VIRGINIA		Vauxhall	32.36	Halifax	25.04
Lynchburg	15.26	BRITISH COLUMBIA		Kentville	27.26

	Average Degree-Days Per Day		Average Degree-Days Per Day		Average Degree-Days Per Day
Sydney	28.31	Ottawa	29.90	Rimouski	33.35
Truro	28.30	St. Catharines	23.29	Ste. Agathe-des-Monts	34.37
Yarmouth	25.43	Sarnia	25.03	St. Ambroise	38.31
ONTARIO		Sault Ste. Marie	32.71	Sept-Iles	38.29
Atikokan	40.18	Thunder Bay	36.73	Sherbrooke	29.95
Belleville	27.18	Toronto	24.02	Victoriaville	31.93
Brampton	27.04	Woodstock	26.83	SASKATCHEWAN	
Chatham	23.32	PRINCE EDWARD ISLAND		Estevan	36.13
Cornwall	28.85	Charlottetown	28.85	Indian Head	38.11
Guelph	27.82	QUEBEC		Melfort	40.88
Hamilton	23.92	Amos	39.85	Moose Jaw	35.24
Harrow	22.97	Berthierville	32.38	Outlook	36.75
Kapuskasing	40.37	Grandes Bergeronnes	35.37	Prince Albert	42.19
Kingston	27.45	L'Assomption	31.49	Regina	38.37
Lindsay	29.45	Lennoxville	31.21	Saskatoon	38.41
London	26.56	Maniwaki	34.26	Scott	40.24
Mount Forest	30.52	Montreal	29.06	Swift Current	34.98
New Liskeard	36.10	Normandin	39.65	Weyburn	35.97
North Bay	32.54	Quebec	31.36	Yorkton	40.19

Note to Table 53: *Derived from Table 15.*

Appendix D

Step One: On the first of the worksheets that follow, fill in the Degree-Day size of your home. Referring to Chapter Thirteen, put down in Column A the month-by-month Degree-Day totals for your home town. Note: If you live in the mountains and your home town is not listed, select the nearest plains city listed and add 90 Degree Days to each month for every 1,000 feet of altitude your home town has more than the listed city.

Step Two: If you are purchasing a properly insulated and designed unit with R-30 insulation, two covers and a horizontal reflector, simply refer to Appendix A to obtain the Btu output, inside your home, from one square foot of collector. Enter those numbers in Column C. If for some reason you are purchasing substandard equipment, refer to Chapter Thirteen. Follow the instructions there to obtain the monthly output numbers, inside your home, from one square foot. Enter those numbers in Column C.

Step Three: Multiply the numbers in Column A by the Degree-Day size of your home, entering the totals, month by month, in Column B.

Step Four: Select, arbitrarily, an area, in square feet, of collector panel that *might* be appropriate for your home. (Tip: Use the Solar-Collector-Gram in Chapter Eleven to guide you.) Enter that number of square feet at the top of Column D, in the space provided. Then multiply that number times the numbers in Column C, entering the results, month by month, in Column D.

Step Five: Multiply the answers in Column D by the number of days in the month, entering the result in Column E.

Step Six: Compare the numbers for each month in Column B and in Column E. Enter the *smaller* of the two numbers in Column F.

Step Seven: Add up Column B, then Column F. Divide the total obtained for Column F by the total obtained for Column B. Multiply that answer by 100, round off, and you will have the percentage of your home heating that the selected size of collector will provide.

Step Eight: If the percentage arrived at is larger than 90%, reduce the square footage of collector for the next trial. Take another worksheet and redo the calculations with the smaller area. If, on the other hand, the percentage obtained in Step Seven is less than you desire, increase the area of collector, take another worksheet and repeat the calculations. NOTE: Do not put on your home a collector that supplies more than 90%.

WORKSHEET FOR ARRIVING AT COLLECTOR NEEDED FOR HOUSE

DEGREE-DAY SIZE OF HOUSE _____

	(A) DEGREE DAYS PER MONTH	(B) B.T.U.'S REQUIRED (DD HOUSE × DD PER MONTH)	(C) B.T.U. OUTPUT IN HOUSE, 1 SQ. FT.	(D) B.T.U.'S PRODUCED BY_____ SQ. FT. OF COLLECTOR		(E) B.T.U. OUTPUT IN HOUSE PER MONTH	(F) USEFUL B.T.U.'S IN HOUSE PER MONTH
SEP					x 30 =		
OCT					x 31 =		
NOV					x 30 =		
DEC					x 31 =		
JAN					x 31 =		
FEB					x 28 =		
MAR					x 31 =		
APR					x 30 =		
MAY					x 31 =		
TOTALS							
		(B)					(F)

F DIVIDED BY B = _____. x 100 = _____ %

WORKSHEET FOR ARRIVING AT COLLECTOR NEEDED FOR HOUSE

DEGREE-DAY SIZE OF HOUSE _____

	(A) DEGREE DAYS PER MONTH	(B) B.T.U.'S REQUIRED (DD HOUSE × DD PER MONTH)	(C) B.T.U. OUTPUT IN HOUSE, 1 SQ. FT.	(D) B.T.U.'S PRODUCED BY_____ SQ. FT. OF COLLECTOR		(E) B.T.U. OUTPUT IN HOUSE PER MONTH	(F) USEFUL B.T.U.'S IN HOUSE PER MONTH
SEP					x 30 =		
OCT					x 31 =		
NOV					x 30 =		
DEC					x 31 =		
JAN					x 31 =		
FEB					x 28 =		
MAR					x 31 =		
APR					x 30 =		
MAY					x 31 =		
TOTALS	(B)						(F)

F DIVIDED BY B = _____. x 100 = _____ %

WORKSHEET FOR ARRIVING AT COLLECTOR NEEDED FOR HOUSE

DEGREE-DAY SIZE OF HOUSE _____

	(A) DEGREE DAYS PER MONTH	(B) B.T.U.'S REQUIRED (DD HOUSE × DD PER MONTH)	(C) B.T.U. OUTPUT IN HOUSE, 1 SQ. FT.	(D) B.T.U.'S PRODUCED BY_____ SQ. FT. OF COLLECTOR		(E) B.T.U. OUTPUT IN HOUSE PER MONTH	(F) USEFUL B.T.U.'S IN HOUSE PER MONTH
SEP					x 30 =		
OCT					x 31 =		
NOV					x 30 =		
DEC					x 31 =		
JAN					x 31 =		
FEB					x 28 =		
MAR					x 31 =		
APR					x 30 =		
MAY					x 31 =		
TOTALS							
		(B)					(F)

F DIVIDED BY B = _____. x 100 = _____ %

WORKSHEET FOR ARRIVING AT COLLECTOR NEEDED FOR HOUSE

DEGREE-DAY SIZE OF HOUSE _____

	(A) DEGREE DAYS PER MONTH	(B) B.T.U.'S REQUIRED (DD HOUSE × DD PER MONTH)	(C) B.T.U. OUTPUT IN HOUSE, 1 SQ. FT.	(D) B.T.U.'S PRODUCED BY_____ SQ. FT. OF COLLECTOR		(E) B.T.U. OUTPUT IN HOUSE PER MONTH	(F) USEFUL B.T.U.'S IN HOUSE PER MONTH
SEP					x 30 =		
OCT					x 31 =		
NOV					x 30 =		
DEC					x 31 =		
JAN					x 31 =		
FEB					x 28 =		
MAR					x 31 =		
APR					x 30 =		
MAY					x 31 =		
TOTALS							
	(B)						(F)

F DIVIDED BY B = _____.   x 100 = _____ %

WORKSHEET FOR ARRIVING AT COLLECTOR NEEDED FOR HOUSE

DEGREE-DAY SIZE OF HOUSE _____

	(A) DEGREE DAYS PER MONTH	(B) B.T.U.'S REQUIRED (DD HOUSE × DD PER MONTH)	(C) B.T.U. OUTPUT IN HOUSE, 1 SQ. FT.	(D) B.T.U.'S PRODUCED BY_____ SQ. FT. OF COLLECTOR		(E) B.T.U. OUTPUT IN HOUSE PER MONTH	(F) USEFUL B.T.U.'S IN HOUSE PER MONTH
SEP					x 30 =		
OCT					x 31 =		
NOV					x 30 =		
DEC					x 31 =		
JAN					x 31 =		
FEB					x 28 =		
MAR					x 31 =		
APR					x 30 =		
MAY					x 31 =		
TOTALS		(B)					(F)

F DIVIDED BY B = _____ x 100 = _____ %

WORKSHEET FOR ARRIVING AT COLLECTOR NEEDED FOR HOUSE

DEGREE-DAY SIZE OF HOUSE _____

	(A) *DEGREE DAYS PER MONTH*	(B) *B.T.U.'S REQUIRED (DD HOUSE × DD PER MONTH)*	(C) *B.T.U. OUTPUT IN HOUSE, 1 SQ. FT.*	(D) *B.T.U.'S PRODUCED BY_____ SQ. FT. OF COLLECTOR*		(E) *B.T.U. OUTPUT IN HOUSE PER MONTH*	(F) *USEFUL B.T.U.'S IN HOUSE PER MONTH*
SEP					x 30 =		
OCT					x 31 =		
NOV					x 30 =		
DEC					x 31 =		
JAN					x 31 =		
FEB					x 28 =		
MAR					x 31 =		
APR					x 30 =		
MAY					x 31 =		
TOTALS		(B)					(F)

F DIVIDED BY B = _____ . x 100 = _____ %

WORKSHEET FOR ARRIVING AT COLLECTOR NEEDED FOR HOUSE

DEGREE-DAY SIZE OF HOUSE _____

	(A) DEGREE DAYS PER MONTH	(B) B.T.U.'S REQUIRED (DD HOUSE × DD PER MONTH)	(C) B.T.U. OUTPUT IN HOUSE, 1 SQ. FT.	(D) B.T.U.'S PRODUCED BY_____ SQ. FT. OF COLLECTOR		(E) B.T.U. OUTPUT IN HOUSE PER MONTH	(F) USEFUL B.T.U.'S IN HOUSE PER MONTH
SEP					x 30 =		
OCT					x 31 =		
NOV					x 30 =		
DEC					x 31 =		
JAN					x 31 =		
FEB					x 28 =		
MAR					x 31 =		
APR					x 30 =		
MAY					x 31 =		
TOTALS		(B)					(F)

F DIVIDED BY B = _____. x 100 = _____ %

Appendix E

Table 54 PERCENTAGE TRANSMISSIVITY OF SINGLE-PANE AND DOUBLE-PANE WINDOW GLASS AT VARIOUS ANGLES OF INCIDENCE

Angle of Incidence Degrees	Percentage Transmitted 1 Pane	2 Pane	Angle of Incidence Degrees	Percentage Transmitted 1 Pane	2 Pane	Angle of Incidence Degrees	Percentage Transmitted 1 Pane	2 Pane
0-20	87.00	87.00	29.00	87.00	76.10	38.00	86.20	75.20
20.25	87.00	76.98	29.25	87.00	76.08	38.25	86.18	75.18
20.50	87.00	76.95	29.50	87.00	76.05	38.50	86.15	75.15
20.75	87.00	76.93	29.75	87.00	76.03	38.75	86.13	75.13
21.00	87.00	76.90	30.00	87.00	76.00	39.00	86.10	75.10
21.25	87.00	76.88	30.25	86.98	75.98	39.25	86.08	75.08
21.50	87.00	76.85	30.50	86.95	75.95	39.50	86.05	75.05
21.75	87.00	76.83	30.75	86.93	75.93	39.75	86.03	75.03
22.00	87.00	76.80	31.00	86.90	75.90	40.00	86.00	75.00
22.25	87.00	76.78	31.25	86.88	75.88	40.25	85.95	74.95
22.50	87.00	76.75	31.50	86.85	75.85	40.50	85.90	74.90
22.75	87.00	76.73	31.75	86.83	75.83	40.75	85.85	74.85
23.00	87.00	76.70	32.00	86.80	75.80	41.00	85.80	74.80
23.25	87.00	76.68	32.25	86.78	75.78	41.25	85.75	74.75
23.50	87.00	76.65	32.50	86.75	75.75	41.50	85.70	74.70
23.75	87.00	76.63	32.75	86.73	75.73	41.75	85.65	74.65
24.00	87.00	76.60	33.00	86.70	75.70	42.00	85.60	74.60
24.25	87.00	76.58	33.25	86.68	75.68	42.25	85.55	74.55
24.50	87.00	76.55	33.50	86.65	75.65	42.50	85.50	74.50
24.75	87.00	76.53	33.75	86.63	75.63	42.75	85.45	74.45
25.00	87.00	76.50	34.00	86.60	75.60	43.00	85.40	74.40
25.25	87.00	76.48	34.25	86.58	75.58	43.25	85.35	74.35
25.50	87.00	76.45	34.50	86.55	75.55	43.50	85.30	74.30
25.75	87.00	76.43	34.75	86.53	75.53	43.75	85.25	74.25
26.00	87.00	76.40	35.00	86.50	75.50	44.00	85.20	74.20
26.25	87.00	76.38	35.25	86.48	75.48	44.25	85.15	74.15
26.50	87.00	76.35	35.50	86.45	75.45	44.50	85.10	74.10
26.75	87.00	76.33	35.75	86.43	75.43	44.75	85.05	74.05
27.00	87.00	76.30	36.00	86.40	75.40	45.00	85.00	74.00
27.25	87.00	76.28	36.25	86.38	75.38	45.25	84.95	73.95
27.50	87.00	76.25	36.50	86.35	75.35	45.50	84.90	73.90
27.75	87.00	76.23	36.75	86.33	75.33	45.75	84.85	73.85
28.00	87.00	76.20	37.00	86.30	75.30	46.00	84.80	73.80
28.25	87.00	76.18	37.25	86.28	75.28	46.25	84.75	73.75
28.50	87.00	76.15	37.50	86.25	75.25	46.50	84.70	73.70
28.75	87.00	76.13	37.75	86.23	75.23	46.75	84.65	73.65

Angle of Incidence Degrees	Percentage Transmitted 1 Pane	2 Pane	Angle of Incidence Degrees	Percentage Transmitted 1 Pane	2 Pane	Angle of Incidence Degrees	Percentage Transmitted 1 Pane	2 Pane
47.00	84.60	73.60	57.00	80.50	68.80	67.00	71.30	57.20
47.25	84.55	73.55	57.25	80.38	68.65	67.25	71.03	56.85
47.50	84.50	73.50	57.50	80.25	68.50	67.50	70.75	56.50
47.75	84.45	73.45	57.75	80.13	68.35	67.75	70.48	56.15
48.00	84.40	73.40	58.00	80.00	68.20	68.00	70.20	55.80
48.25	84.35	73.35	58.25	79.88	68.05	68.25	69.93	55.45
48.50	84.30	73.30	58.50	79.75	67.90	68.50	69.65	55.10
48.75	84.25	73.25	58.75	79.63	67.75	68.75	69.38	54.75
49.00	84.20	73.20	59.00	79.50	67.60	69.00	69.10	54.40
49.25	84.15	73.15	59.25	79.38	67.45	69.25	68.83	54.05
49.50	84.10	73.10	59.50	79.25	67.30	69.50	68.55	53.70
49.75	84.05	73.05	59.75	79.13	67.15	69.75	68.28	53.35
50.00	84.00	73.00	60.00	79.00	67.00	70.00	68.00	53.00
50.25	83.88	72.85	60.25	78.73	66.65	70.25	67.35	52.30
50.50	83.75	72.70	60.50	78.45	66.30	70.50	66.70	51.60
50.75	83.63	72.55	60.75	78.18	65.95	70.75	66.05	50.90
51.00	83.50	72.40	61.00	77.90	65.60	71.00	65.40	50.20
51.25	83.38	72.25	61.25	77.63	65.25	71.25	64.75	49.50
51.50	83.25	72.10	61.50	77.35	64.90	71.50	64.10	48.80
51.75	83.13	71.95	61.75	77.08	64.55	71.75	63.45	48.10
52.00	83.00	71.80	62.00	76.80	64.20	72.00	62.80	47.40
52.25	82.88	71.65	62.25	76.53	63.85	72.25	62.15	46.70
52.50	82.75	71.50	62.50	76.25	63.50	72.50	61.50	46.00
52.75	82.63	71.35	62.75	75.98	63.15	72.75	60.85	45.30
53.00	82.50	71.20	63.00	75.70	62.80	73.00	60.20	44.60
53.25	82.38	71.05	63.25	75.43	62.45	73.25	59.55	43.90
53.50	82.25	70.90	63.50	75.15	62.10	73.50	58.90	43.20
53.75	82.13	70.75	63.75	74.88	61.75	73.75	58.25	42.50
54.00	82.00	70.60	64.00	74.60	61.40	74.00	57.60	41.80
54.25	81.88	70.45	64.25	74.33	61.05	74.25	56.95	41.10
54.50	81.75	70.30	64.50	74.05	60.70	74.50	56.30	40.40
54.75	81.63	70.15	64.75	73.78	60.35	74.75	55.65	39.70
55.00	81.50	70.00	65.00	73.50	60.00	75.00	55.00	39.00
55.25	81.38	69.85	65.25	73.23	59.65	75.25	54.35	38.30
55.50	81.25	69.70	65.50	72.95	59.30	75.50	53.70	37.60
55.75	81.13	69.55	65.75	72.68	58.95	75.75	53.05	36.90
56.00	81.00	69.40	66.00	72.40	58.60	76.00	52.40	36.20
56.25	80.88	69.25	66.25	72.13	58.25	76.25	51.75	35.50
56.50	80.75	69.10	66.50	71.85	57.90	76.50	51.10	34.80
56.75	80.63	68.95	66.75	71.58	57.55	76.75	50.45	34.10

Angle of Incidence Degrees	Percentage Transmitted		Angle of Incidence Degrees	Percentage Transmitted		Angle of Incidence Degrees	Percentage Transmitted	
	1 Pane	2 Pane		1 Pane	2 Pane		1 Pane	2 Pane
77.00	49.80	33.40	80.75	38.06	22.66	84.50	18.38	10.94
77.25	49.15	32.70	81.00	36.75	21.88	84.75	17.06	10.16
77.50	48.50	32.00	81.25	35.44	21.09	85.00	15.75	9.38
77.75	47.85	31.30	81.50	34.13	20.31	85.25	14.44	8.50
78.00	47.20	30.60	81.75	32.81	19.53	85.50	13.13	7.81
78.25	46.55	29.90	82.00	31.50	18.75	85.75	11.81	7.03
78.50	45.90	29.20	82.25	30.19	17.97	86.00	10.50	6.25
78.75	45.25	28.50	82.50	28.33	17.19	86.25	9.19	5.47
79.00	44.60	27.80	82.75	27.56	16.41	86.50	7.88	4.69
79.25	43.95	27.10	83.00	26.25	15.63	86.75	6.56	3.91
79.50	43.30	26.40	83.25	24.94	14.84	87.00	5.25	3.13
79.75	42.65	25.70	83.50	23.63	14.06	87.25	3.94	2.35
80.00	42.00	25.00	83.75	22.31	13.28	87.50	2.63	1.56
80.25	40.69	24.22	84.00	21.00	12.50	87.75	1.31	0.78
80.50	39.38	23.44	84.25	19.69	11.72	88.00	0	0

Note to Transmissivity Table: *Adapted from* A.S.H.R.A.E. Handbook and Product Directory, *Table 4, "Variation with Incident Angle of Transmittance for Single and Double Glazing and Absorptance for Flat Black Paint," p. 59.6, New York, 1974; and author's transmittance measurements.*

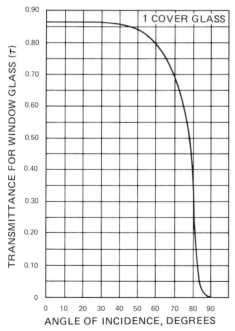

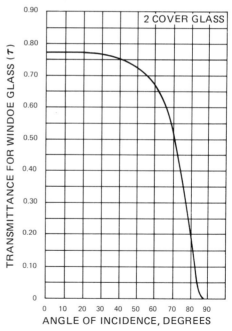

NOTE: For Water-White Glass, Multiply by 1.03

FIGURE 7: Graphs show transmittance of direct solar radiation through cover(s) on solar collector.

Appendix F
CONVERSIONS AND CONSTANTS

Table 55 **CONVERTING FAHRENHEIT DEGREE DAYS (BASE 65° F) TO CELSIUS DEGREE DAYS (BASE 18° C)**

Instructions: To convert a monthly value of Fahrenheit Degree Days to a monthly value of Celsius Degree Days, first find the digits preceding the final digit in the monthly value of Fahrenheit Degree Days in the vertical "F" column, below. Then find the final digit at the top of the page in the horizontal "C" row. At the point where lines drawn from the two would intersect is the number of Celsius Degree Days equivalent to that Fahrenheit Degree-Days number.

Example: Find the Celsius Degree Days equivalent to a monthly value of 995 Fahrenheit Degree Days. In the "F" column, you would find 99 (the digits preceding the final digit). At the top of the page, you would find the numeral 5. The two intersect at 543. Thus, 543 Celsius Degree Days are equivalent to 995 Fahrenheit Degree Days.

DEGREE DAYS BELOW 65 ° F TO DEGREE DAYS BELOW 18 ° C

F	C 0	1	2	3	4	5	6	7	8	9
0	0	1	1	2	2	2	3	3	4	4
1	5	6	6	6	7	7	8	8	9	9
2	9	10	11	11	12	12	13	13	14	14
3	14	15	16	16	17	17	18	18	18	19
4	19	20	21	21	22	22	23	23	23	24
5	24	25	26	26	27	27	28	28	29	29
6	29	30	31	31	32	32	33	33	34	34
7	35	35	36	36	37	37	38	38	39	39
8	40	41	41	42	42	42	43	43	44	44
9	45	46	46	47	47	48	48	49	49	49
10	50	51	51	52	52	53	53	54	54	55
11	56	56	57	57	58	58	59	59	60	61
12	61	61	62	62	63	63	64	64	65	66
13	66	67	67	67	68	68	69	69	70	71
14	71	72	72	73	73	74	74	75	76	76
15	77	77	78	78	79	79	79	80	81	81
16	82	82	83	83	84	84	85	86	86	87
17	87	88	88	89	89	90	91	91	92	92
18	93	93	93	94	94	95	96	96	97	97
19	98	98	99	99	100	101	101	102	102	103
20	103	104	104	105	106	106	107	107	108	108
21	109	109	110	111	111	111	112	112	113	113
22	114	114	115	116	116	117	117	118	118	119
23	119	120	121	121	122	122	123	123	124	124
24	125	126	126	127	127	128	128	129	129	130
25	131	131	132	132	133	133	134	134	134	135
26	136	136	137	137	138	138	139	139	140	141
27	141	142	142	143	143	144	144	145	146	146
28	147	147	148	148	149	149	150	151	151	152
29	152	153	153	154	154	155	155	156	157	157

Table 55 233

DEGREE DAYS BELOW 65° F TO DEGREE DAYS BELOW 18°C

F	0	1	2	3	4	5	6	7	8	9
30	158	158	159	159	160	161	161	162	162	163
31	163	164	164	165	166	166	167	167	168	168
32	169	169	170	171	171	171	172	172	173	173
33	174	174	175	176	176	177	177	178	178	179
34	179	180	181	181	182	182	183	183	184	184
35	185	186	186	187	187	188	188	189	189	190
36	191	191	192	192	193	193	194	194	195	196
37	196	197	197	198	198	199	199	200	201	201
38	202	202	203	203	203	204	205	206	206	207
39	207	208	208	209	209	210	211	211	212	212
40	213	213	214	214	215	216	216	217	217	218
41	218	219	219	220	221	221	222	222	223	223
42	224	224	225	226	226	227	227	228	228	229
43	229	230	231	231	232	232	233	233	234	234
44	235	236	236	237	237	238	238	239	239	240
45	241	241	242	242	243	243	244	244	245	246
46	246	247	247	248	248	249	249	249	250	251
47	251	252	252	253	253	254	254	255	256	256
48	257	257	258	258	259	259	260	261	261	262
49	262	263	263	264	264	265	266	266	267	267
50	268	268	269	269	270	271	271	272	272	273
51	273	274	274	275	276	276	277	277	278	278
52	279	279	280	281	281	282	282	283	283	284
53	284	285	286	286	287	287	288	288	289	289
54	290	291	291	292	292	293	293	294	294	295
55	296	296	297	297	298	298	299	299	300	301
56	301	302	302	303	303	304	304	305	306	306
57	307	307	308	308	309	309	310	311	311	312
58	312	313	313	314	314	315	316	316	317	317
59	318	318	319	319	320	321	321	322	322	323
60	323	324	324	325	326	326	327	327	328	328
61	329	329	330	331	331	332	332	333	333	334
62	334	335	336	336	337	337	338	338	339	339
63	340	341	341	342	342	343	343	344	344	345
64	346	346	347	347	348	348	349	349	350	351
65	351	352	352	353	353	354	354	355	356	356
66	357	357	358	358	359	359	360	361	361	362
67	362	363	363	364	364	365	366	366	367	367
68	368	368	369	369	370	371	371	372	372	373
69	373	374	374	375	376	376	377	377	378	378

DEGREE DAYS BELOW 65°F TO DEGREE DAYS BELOW 18°C

C

F	0	1	2	3	4	5	6	7	8	9
70	379	379	380	381	381	382	382	383	383	384
71	384	385	386	386	387	387	388	388	389	389
72	390	391	391	392	392	393	393	394	394	395
73	396	396	397	397	398	398	399	399	400	401
74	401	402	402	403	403	404	404	405	406	406
75	407	407	408	408	409	409	410	411	411	412
76	412	413	413	414	414	415	416	416	417	417
77	418	418	419	419	420	421	421	422	422	423
78	423	424	424	425	426	426	427	427	428	428
79	429	429	430	431	431	432	432	433	433	434
80	434	435	436	436	437	437	438	438	439	439
81	440	441	441	442	442	443	443	444	444	445
82	446	446	447	447	448	448	449	449	450	451
83	451	452	452	453	453	454	454	455	456	456
84	457	457	458	458	459	459	460	461	461	462
85	462	463	463	464	464	465	466	466	467	467
86	468	468	469	469	470	471	471	472	472	473
87	473	474	474	475	476	476	477	477	478	478
88	479	479	480	481	481	482	482	483	483	484
89	484	485	486	486	487	487	488	488	489	489
90	490	491	491	492	492	493	493	494	494	495
91	496	496	497	497	498	498	499	499	500	501
92	501	502	502	503	503	504	504	505	506	506
93	507	507	508	508	509	509	510	511	511	512
94	512	513	513	514	514	515	516	516	517	517
95	518	518	519	519	520	521	521	522	522	523
96	523	524	524	525	526	526	527	527	528	528
97	529	529	530	531	531	532	532	533	533	534
98	534	535	536	536	537	537	538	538	539	539
99	540	541	541	542	542	543	543	544	544	545
100	546	546	547	547	548	548	549	549	550	551
101	551	552	552	553	553	554	554	555	556	556
102	557	557	558	558	559	559	560	561	561	562
103	562	563	563	564	564	565	566	566	567	567
104	568	568	569	569	570	571	571	572	572	573
105	573	574	574	575	576	576	577	577	578	578
106	579	579	580	581	581	582	582	583	583	584
107	584	585	586	586	587	587	588	588	589	589
108	590	591	591	592	592	593	593	594	594	595
109	596	596	597	597	598	598	599	599	600	601

Table 55 235

DEGREE DAYS BELOW 65°F TO DEGREE DAYS BELOW 18°C

C

F	0	1	2	3	4	5	6	7	8	9
110	601	602	602	603	603	604	604	605	606	606
111	607	607	608	608	609	609	610	611	611	612
112	612	613	613	614	614	615	616	616	617	617
113	618	618	619	619	620	621	621	622	622	623
114	623	624	624	625	626	626	627	627	628	628
115	629	629	630	631	631	632	632	633	633	634
116	634	635	636	636	637	637	638	638	639	639
117	640	641	641	642	642	643	643	644	644	645
118	646	646	647	647	648	648	649	649	650	651
119	651	652	652	653	653	654	654	655	656	656
120	657	657	658	658	659	659	660	661	661	662
121	662	663	663	664	664	665	666	666	667	667
122	668	668	669	669	670	671	671	672	672	673
123	673	674	674	675	676	676	677	677	678	678
124	679	679	680	681	681	682	682	683	683	684
125	684	685	686	686	687	687	688	688	689	689
126	690	691	691	692	692	693	693	694	694	695
127	696	696	697	697	698	698	699	699	700	701
128	701	702	702	703	703	704	704	705	706	706
129	707	707	708	708	709	709	710	711	711	712
130	712	713	713	714	714	715	716	716	717	717
131	718	718	719	719	720	721	721	722	722	723
132	723	724	724	725	726	726	727	727	728	728
133	729	729	730	731	731	732	732	733	733	734
134	734	735	736	736	737	737	738	738	739	739
135	740	741	741	742	742	743	743	744	744	745
136	746	746	747	747	748	748	749	749	750	751
137	751	752	752	753	753	754	754	755	756	756
138	757	757	758	758	759	759	760	761	761	762
139	762	763	763	764	764	765	766	766	767	767
140	768	768	769	769	770	771	771	772	772	773
141	773	774	774	775	776	776	777	777	778	778
142	779	779	780	781	781	782	782	783	783	784
143	784	785	786	786	787	787	788	788	789	789
144	790	791	791	792	792	793	793	794	794	795
145	796	796	797	797	798	798	799	799	800	801
146	801	802	802	803	803	804	804	805	806	806
147	807	807	808	808	809	809	810	811	811	812
148	812	813	813	814	814	815	816	816	817	817
149	818	818	819	819	820	821	821	822	822	823

DEGREE DAYS BELOW 65°F TO DEGREE DAYS BELOW 18°C

C

F	*0*	*1*	*2*	*3*	*4*	*5*	*6*	*7*	*8*	*9*
150	823	824	824	825	826	826	827	827	828	828
151	829	829	830	831	831	832	832	833	833	834
152	834	835	836	836	837	837	838	838	839	839
153	840	841	841	842	842	843	843	844	844	845
154	846	846	847	847	848	848	849	849	850	851
155	851	852	852	853	853	854	854	855	856	856
156	857	857	858	858	859	859	860	861	861	862
157	862	863	863	864	864	865	866	866	867	867
158	868	868	869	869	870	871	871	872	872	873
159	873	874	874	875	876	876	877	877	878	878
160	879	879	880	881	881	882	882	883	883	884
161	884	885	886	886	887	887	888	888	889	889
162	890	891	891	892	892	893	893	894	894	895
163	896	896	897	897	898	898	899	899	900	901
164	901	902	902	903	903	904	904	905	906	906
165	907	907	908	908	909	909	910	911	911	912
166	912	913	913	914	914	915	916	916	917	917
167	918	918	919	919	920	921	921	922	922	923
168	923	924	924	925	926	926	927	927	928	928
169	929	929	930	931	931	932	932	933	933	934
170	934	935	936	936	937	937	938	938	939	939
171	940	941	941	942	942	943	943	944	944	945
172	946	946	947	947	948	948	949	949	950	951
173	951	952	952	953	953	954	954	955	956	956
174	957	957	958	958	959	959	960	961	961	962
175	962	963	963	964	964	965	966	966	967	967
176	968	968	969	969	970	971	971	972	972	973
177	973	974	974	975	976	976	977	977	978	978
178	979	979	980	981	981	982	982	983	983	984
179	984	985	986	986	987	987	988	988	989	989
180	990	991	991	992	992	993	993	994	994	995
181	996	996	997	997	998	998	999	999	1,000	1,001
182	1,001	1,002	1,002	1,003	1,003	1,004	1,004	1,005	1,006	1,006
183	1,007	1,007	1,008	1,008	1,009	1,009	1,010	1,011	1,011	1,012
184	1,012	1,013	1,013	1,014	1,014	1,015	1,016	1,016	1,017	1,017
185	1,018	1,018	1,019	1,019	1,020	1,021	1,021	1,022	1,022	1,023
186	1,023	1,024	1,024	1,025	1,026	1,026	1,027	1,027	1,028	1,028
187	1,029	1,029	1,030	1,031	1,031	1,032	1,032	1,033	1,033	1,034
188	1,034	1,035	1,036	1,036	1,037	1,037	1,038	1,038	1,039	1,039
189	1,040	1,041	1,041	1,042	1,042	1,043	1,043	1,044	1,044	1,045

Table 55 237

DEGREE DAYS BELOW 65 °F TO DEGREE DAYS BELOW 18 °C

C

F	0	1	2	3	4	5	6	7	8	9
190	1,046	1,046	1,047	1,047	1,048	1,048	1,049	1,049	1,050	1,051
191	1,051	1,052	1,052	1,053	1,053	1,054	1,054	1,055	1,056	1,056
192	1,057	1,057	1,058	1,058	1,059	1,059	1,060	1,061	1,061	1,062
193	1,062	1,063	1,063	1,064	1,064	1,065	1,066	1,066	1,067	1,067
194	1,068	1,068	1,069	1,069	1,070	1,071	1,071	1,072	1,072	1,073
195	1,073	1,074	1,074	1,075	1,076	1,076	1,077	1,077	1,078	1,078
196	1,079	1,079	1,080	1,081	1,081	1,082	1,082	1,083	1,083	1,084
197	1,084	1,085	1,086	1,086	1,087	1,087	1,088	1,088	1,089	1,089
198	1,090	1,091	1,091	1,092	1,092	1,093	1,093	1,094	1,094	1,095
199	1,096	1,096	1,097	1,097	1,098	1,098	1,099	1,099	1,100	1,101
200	1,101	1,102	1,102	1,103	1,103	1,104	1,104	1,105	1,106	1,106
201	1,107	1,107	1,108	1,108	1,109	1,109	1,110	1,111	1,111	1,112
202	1,112	1,113	1,113	1,114	1,114	1,115	1,116	1,116	1,117	1,117
203	1,118	1,118	1,119	1,119	1,120	1,121	1,121	1,122	1,122	1,123
204	1,123	1,124	1,124	1,125	1,126	1,126	1,127	1,127	1,128	1,128
205	1,129	1,129	1,130	1,131	1,131	1,132	1,132	1,133	1,133	1,134
206	1,134	1,135	1,136	1,136	1,137	1,137	1,138	1,138	1,139	1,139
207	1,140	1,141	1,141	1,142	1,142	1,143	1,143	1,144	1,144	1,145
208	1,146	1,146	1,147	1,147	1,148	1,148	1,149	1,149	1,150	1,151
209	1,151	1,152	1,152	1,153	1,153	1,154	1,154	1,155	1,156	1,156
210	1,157	1,157	1,158	1,158	1,159	1,159	1,160	1,161	1,161	1,162
211	1,162	1,163	1,163	1,164	1,164	1,165	1,166	1,166	1,167	1,167
212	1,168	1,168	1,169	1,169	1,170	1,171	1,171	1,172	1,172	1,173
213	1,173	1,174	1,174	1,175	1,176	1,176	1,177	1,177	1,178	1,178
214	1,179	1,179	1,180	1,181	1,181	1,182	1,182	1,183	1,183	1,184
215	1,184	1,185	1,186	1,186	1,187	1,187	1,188	1,188	1,189	1,189
216	1,190	1,191	1,191	1,192	1,192	1,193	1,193	1,194	1,194	1,195
217	1,196	1,196	1,197	1,197	1,198	1,198	1,199	1,199	1,200	1,201
218	1,201	1,202	1,202	1,203	1,203	1,204	1,204	1,205	1,206	1,206
219	1,207	1,207	1,208	1,208	1,209	1,209	1,210	1,211	1,211	1,212
220	1,212	1,213	1,213	1,214	1,214	1,215	1,216	1,216	1,217	1,217
221	1,218	1,218	1,219	1,219	1,220	1,221	1,221	1,222	1,222	1,223
222	1,223	1,224	1,224	1,225	1,226	1,226	1,227	1,227	1,228	1,228
223	1,229	1,229	1,230	1,231	1,231	1,232	1,232	1,233	1,233	1,234
224	1,234	1,235	1,236	1,236	1,237	1,237	1,238	1,238	1,239	1,239
225	1,240	1,241	1,241	1,242	1,242	1,243	1,243	1,244	1,244	1,245
226	1,246	1,246	1,247	1,247	1,248	1,248	1,249	1,249	1,250	1,251
227	1,251	1,252	1,252	1,253	1,253	1,254	1,254	1,255	1,256	1,256
228	1,257	1,257	1,258	1,258	1,259	1,259	1,260	1,261	1,261	1,262
229	1,262	1,263	1,263	1,264	1,264	1,265	1,266	1,266	1,267	1,267

DEGREE DAYS BELOW 65°F TO DEGREE DAYS BELOW 18°C

C

F	0	1	2	3	4	5	6	7	8	9
230	1,268	1,268	1,269	1,269	1,270	1,271	1,271	1,272	1,272	1,273
231	1,273	1,274	1,274	1,275	1,276	1,276	1,277	1,277	1,278	1,278
232	1,279	1,279	1,280	1,281	1,281	1,282	1,282	1,283	1,283	1,284
233	1,284	1,285	1,286	1,286	1,287	1,287	1,288	1,288	1,289	1,289
234	1,290	1,291	1,291	1,292	1,292	1,293	1,293	1,294	1,294	1,295
235	1,296	1,296	1,297	1,297	1,298	1,298	1,299	1,299	1,300	1,301
236	1,301	1,302	1,302	1,303	1,303	1,304	1,304	1,305	1,306	1,306
237	1,307	1,307	1,308	1,308	1,309	1,309	1,310	1,311	1,311	1,312
238	1,312	1,313	1,313	1,314	1,314	1,315	1,316	1,316	1,317	1,317
239	1,318	1,318	1,319	1,319	1,320	1,321	1,321	1,322	1,322	1,323
240	1,323	1,324	1,324	1,325	1,326	1,326	1,327	1,327	1,328	1,328
241	1,329	1,329	1,330	1,331	1,331	1,332	1,332	1,333	1,333	1,334
242	1,334	1,335	1,336	1,336	1,337	1,337	1,338	1,338	1,339	1,339
243	1,340	1,341	1,341	1,342	1,342	1,343	1,343	1,344	1,344	1,345
244	1,346	1,346	1,347	1,347	1,348	1,348	1,349	1,349	1,350	1,351
245	1,351	1,352	1,352	1,353	1,353	1,354	1,354	1,355	1,356	1,356
246	1,357	1,357	1,358	1,358	1,359	1,359	1,360	1,361	1,361	1,362
247	1,362	1,363	1,363	1,364	1,364	1,365	1,366	1,366	1,367	1,367
248	1,368	1,368	1,369	1,369	1,370	1,371	1,371	1,372	1,372	1,373
249	1,373	1,374	1,374	1,375	1,376	1,376	1,377	1,377	1,378	1,378
250	1,379	1,379	1,380	1,381	1,381	1,382	1,382	1,383	1,383	1,384
251	1,384	1,385	1,386	1,386	1,387	1,387	1,388	1,388	1,389	1,389
252	1,390	1,391	1,391	1,392	1,392	1,393	1,393	1,394	1,394	1,395
253	1,396	1,396	1,397	1,397	1,398	1,398	1,399	1,399	1,400	1,401
254	1,401	1,402	1,402	1,403	1,403	1,404	1,404	1,405	1,406	1,406
255	1,407	1,407	1,408	1,408	1,409	1,409	1,410	1,411	1,411	1,412
256	1,412	1,413	1,413	1,414	1,414	1,415	1,416	1,416	1,417	1,417
257	1,418	1,418	1,419	1,419	1,420	1,421	1 421	1,422	1,422	1,423
258	1,423	1,424	1,424	1,425	1,426	1,426	1,427	1,427	1,428	1,428
259	1,429	1,429	1,430	1,431	1,431	1,432	1,432	1,433	1,433	1,434
260	1,434	1,435	1,436	1,436	1,437	1,437	1,438	1,438	1,439	1,439
261	1,440	1,441	1,441	1,442	1,442	1,443	1,443	1,444	1,444	1,445
262	1,446	1,446	1,447	1,447	1,448	1,448	1,449	1,449	1,450	1.451
263	1,451	1,452	1,452	1,453	1,453	1,454	1,454	1,455	1,456	1,456
264	1,457	1,457	1,458	1,458	1,459	1,459	1,460	1,461	1,461	1,462
265	1,462	1,463	1,463	1,464	1,464	1,465	1,466	1,466	1,467	1,467
266	1,468	1,468	1,469	1,469	1,470	1,471	1,471	1,472	1,472	1,473
267	1,473	1,474	1,474	1,475	1,476	1,476	1,477	1,477	1,478	1,478
268	1,479	1,479	1,480	1,481	1,481	1,482	1,482	1,483	1,483	1,484
269	1,484	1,485	1,486	1,486	1,487	1,487	1,488	1,488	1,489	1,489

Table 55 239

DEGREE DAYS BELOW 65°F TO DEGREE DAYS BELOW 18°C

C

F	0	1	2	3	4	5	6	7	8	9
270	1,490	1,491	1,491	1,492	1,492	1,493	1,493	1,494	1,494	1,495
271	1,496	1,496	1,497	1,497	1,498	1,498	1,499	1,499	1,500	1,501
272	1,501	1,502	1,502	1,503	1,503	1,504	1,504	1,505	1,506	1,506
273	1,507	1,507	1,508	1,508	1,509	1,509	1,510	1,511	1,511	1,512
274	1,512	1,513	1,513	1,514	1,514	1,515	1,516	1,516	1,517	1,517
275	1,518	1,518	1,519	1,519	1,520	1,521	1,521	1,522	1,522	1,523
276	1,523	1,524	1,524	1,525	1,526	1,526	1,527	1,527	1,528	1,528
277	1,529	1,529	1,530	1,531	1,531	1 532	1,532	1,533	1,533	1,534
278	1,534	1,535	1,536	1,536	1,537	1,537	1,538	1,538	1,539	1,539
279	1,540	1,541	1,541	1,542	1,542	1,543	1,543	1,544	1,544	1,545
280	1,546	1,546	1,547	1,547	1,548	1,548	1,549	1,549	1,550	1,551
281	1,551	1,552	1,552	1,553	1,553	1,554	1,554	1,555	1,556	1,556
282	1,557	1,557	1,558	1,558	1,559	1,559	1,560	1,561	1,561	1,562
283	1,562	1,563	1,563	1,564	1,564	1,565	1,566	1,566	1,567	1,567
284	1,568	1,568	1,569	1,569	1,570	1,571	1,571	1,572	1,572	1,573
285	1,573	1,574	1,574	1,575	1,576	1,576	1,577	1,577	1,578	1,578
286	1,579	1,579	1,580	1,581	1,581	1,582	1,582	1,583	1,583	1,584
287	1,584	1,585	1,586	1,586	1,587	1,587	1,588	1,588	1,589	1,589
288	1,590	1,591	1,591	1,592	1,592	1,593	1,593	1,594	1,594	1,595
289	1,596	1,596	1,597	1,597	1,598	1,598	1,599	1,599	1,600	1,601
290	1,601	1,602	1,602	1,603	1,603	1,604	1,604	1,605	1,606	1,606
291	1,607	1,607	1,608	1,608	1,609	1,609	1,610	1,611	1,611	1,612
292	1,612	1,613	1,613	1,614	1,614	1,615	1,616	1,616	1,617	1,617
293	1,618	1,618	1,619	1,619	1,620	1,621	1,621	1,622	1,622	1,623
294	1,623	1,624	1,624	1,625	1,626	1,626	1,627	1,627	1,628	1,628
295	1,629	1,629	1,630	1,631	1,631	1,632	1,632	1,633	1,633	1,634
296	1,634	1,635	1,636	1,636	1,637	1,637	1,638	1,638	1,639	1,639
297	1,640	1,641	1,641	1,642	1,642	1,643	1,643	1,644	1,644	1,645
298	1,646	1,646	1,647	1,647	1,648	1,648	1,649	1,649	1,650	1,651
299	1,651	1,652	1,652	1,653	1,653	1,654	1,654	1,655	1,656	1,656
300	1,657	1,657	1,658	1,658	1,659	1,659	1,660	1,661	1,661	1,662
301	1,662	1,663	1,663	1,664	1,664	1,665	1,666	1,666	1,667	1,667
302	1,668	1,668	1,669	1,669	1,670	1,671	1,671	1,672	1,672	1,673
303	1,673	1,674	1,674	1,675	1,676	1,676	1,677	1,677	1,678	1,678
304	1,679	1,679	1,680	1,681	1,681	1,682	1,682	1,683	1,683	1,684
305	1,684	1,685	1,686	1,686	1,687	1,687	1,688	1,688	1,689	1,689
306	1,690	1,691	1,691	1,692	1,692	1,693	1,693	1,694	1,694	1,695
307	1,696	1,696	1,697	1,697	1,698	1,698	1,699	1,699	1,700	1,701
308	1,701	1,702	1,702	1,703	1,703	1,704	1,704	1,705	1,706	1,706
309	1,707	1,707	1,708	1,708	1,709	1,709	1,710	1,711	1,711	1,712

DEGREE DAYS BELOW 65°F TO DEGREE DAYS BELOW 18°C

C

F	0	1	2	3	4	5	6	7	8	9
310	1,712	1,713	1,713	1,714	1,714	1,715	1,716	1,716	1,717	1,717
311	1,718	1,718	1,719	1,719	1,720	1,721	1,721	1,722	1,722	1,723
312	1,723	1,724	1,724	1,725	1,726	1,726	1,727	1,727	1,728	1,728
313	1,729	1,729	1,730	1,731	1,731	1,732	1,732	1,733	1,733	1,734
314	1,734	1,735	1,736	1,736	1,737	1,737	1,738	1,738	1,739	1,739
315	1,740	1,741	1,741	1,742	1,742	1,743	1,743	1,744	1,744	1,745
316	1,746	1,746	1,747	1,747	1,748	1,748	1,749	1,749	1,750	1,751
317	1,751	1,752	1,752	1,753	1,753	1,754	1,754	1,755	1,756	1,756
318	1,757	1,757	1,758	1,758	1,759	1,759	1,760	1,761	1,761	1,762
319	1,762	1,763	1,763	1,764	1,764	1,765	1,766	1,766	1,767	1,767
320	1,768	1,768	1,769	1,769	1,770	1,771	1,771	1,772	1,772	1,773
321	1,773	1,774	1,774	1,775	1,776	1,776	1,777	1,777	1,778	1,778
322	1,779	1,779	1,780	1,781	1,781	1,782	1,782	1,783	1,783	1,784
323	1,784	1,785	1,786	1,786	1,787	1,787	1,788	1,788	1,789	1,789
324	1,790	1,791	1,791	1,792	1,792	1,793	1,793	1,794	1,794	1,795
325	1,796	1,796	1,797	1,797	1,798	1,798	1,799	1,799	1,800	1,801
326	1,801	1,802	1,802	1,803	1,803	1,804	1,804	1,805	1,806	1,806
327	1,807	1,807	1,808	1,808	1,809	1,809	1,810	1,811	1,811	1,812
328	1,812	1,813	1,813	1,814	1,814	1,815	1,816	1,816	1,817	1,817
329	1,818	1,818	1,819	1,819	1,820	1,821	1,821	1,822	1,822	1,823
330	1,823	1,824	1,824	1,825	1,826	1,826	1,827	1,827	1,828	1,828
331	1,829	1,829	1,830	1,831	1,831	1,832	1,832	1,833	1,833	1,834
332	1,834	1,835	1,836	1,836	1,837	1,837	1,838	1,838	1,839	1,839
333	1,840	1,841	1,841	1,842	1,842	1,843	1,843	1,844	1,844	1,845
334	1,846	1,846	1,847	1,847	1,848	1,848	1,849	1,849	1,850	1,851
335	1,851	1,852	1,852	1,853	1,853	1,854	1,854	1,855	1,856	1,856
336	1,857	1,857	1,858	1,858	1,859	1,859	1,860	1,861	1,861	1,862
337	1,862	1,863	1,863	1,864	1,864	1,865	1,866	1,866	1,867	1,867
338	1,868	1,868	1,869	1,869	1,870	1,871	1,871	1,872	1,872	1,873
339	1,873	1,874	1,874	1,875	1,876	1,876	1,877	1,877	1,878	1,878
340	1,879	1,879	1,880	1,881	1,881	1,882	1,882	1,883	1,883	1,884
341	1,884	1,885	1,886	1,886	1,887	1,887	1,888	1,888	1,889	1,889
342	1,890	1,891	1,891	1,892	1,892	1,893	1,893	1,894	1,894	1,895
343	1,896	1,896	1,897	1,897	1,898	1,898	1,899	1,899	1,900	1,901
344	1,901	1,902	1,902	1,903	1,903	1,904	1,904	1,905	1,906	1,906
345	1,907	1,907	1,908	1,908	1,909	1,909	1,910	1,911	1,911	1,912
346	1,912	1,913	1,913	1,914	1,914	1,915	1,916	1,916	1,917	1,917
347	1,918	1,918	1,919	1,919	1,920	1,921	1,921	1,922	1,922	1,923
348	1,923	1,924	1,924	1,925	1,926	1,926	1,927	1,927	1,928	1,928
349	1,929	1,929	1,930	1,931	1,931	1,932	1,932	1,933	1,933	1,934
350	1,934									

Note to Table 55: *Table extracted from "Normals of Heating Degree-Days Below 18°C for Stations Listed in the 'Monthly Summary of Degree-Days'," by Boyd, D.W., Meteorological Applications Branch, Atmospheric Environment Service, Department of the Environment, Downsview, Ontario, Canada, 1975.*

Table 56 CONVERSION FACTORS

Length	1 inch = 0.0833 ft		1 watt/cm^2
	= 2.54 cm		= 3,170 Btu/ft$^2 \cdot$ hr
	1 micron = 3.281 10^{-8} ft		1 langley/day
	= 10^{-4} cm		= 3.687 Btu/ft$^2 \cdot$ day
	1 angstrom unit = 10^{-8} cm	*Pressure*	1 in. of water
			= 0.03613 lb/in^2
Heat/Power/Work	1 Btu = 778.3 ft $\cdot$ lb		1 ft of water
	= 252 cal		= 62.43 lb/ft^2
	1 kw $\cdot$ hr = 3,413 Btu		
	1 hp = 2,544 Btu/hr	*Thermal Conductivity*	1 cal/cm $\cdot$ sec $\cdot$ °C
	1 ton refrigeration		= 241.9 Btu/ft $\cdot$ hr $\cdot$ °F
	= 12,000 Btu/hr		1 watt/cm $\cdot$ °C
Heat Flux	1 langley = 1 cal/cm^2		= 57.79 Btu/ft $\cdot$ hr $\cdot$ °F
	1 cal/cm$^2 \cdot$ min	*Volume*	1 ft^3 = 1,728 in^3 = 62.43 lb
	= 221.2 Btu/ft$^2 \cdot$ hr		water = 7.481 gal (U.S.)

Table 57 MISCELLANEOUS CONSTANTS

1 solar constant	2.0 langleys/min	1 degree of latitude at 40° North = 69 miles
	442 Btu/ft$^2 \cdot$ hr	Acceleration due to gravity at sea level
	= 129.5 watts/ft^2	= 32.17 ft/sec^2
	0.1394 watts/cm^2	Atmospheric pressure at sea level = 14.7 lb/in^2
	= 1.394 kw/m^2	Density of air at sea level, 60° F = 0.075 lb/ft^3

Appendix G DECEPTIVE PRACTICES

This list of prevalent deceptive practices in the infant solar industry may prove useful to you in evaluating both the solar dealer and the solar equipment you are considering.

(1) Selling solar collector panels with only one cover in northern latitudes (north of 30 degrees).

(2) Selling solar collector units with only one, two or three inches of fibrous-glass batt insulation (less than in the walls of most homes!)

(3) Giving the consumer the impression that he or she can easily engineer storage and fluid systems and controls to be compatible with the sunlight-using system that is being sold.

(4) Not telling the customer the Btu output, inside the home, of the entire pre-engineered solar heating system.

(5) Not telling the customer about the maintenance costs of the collector/distribution system.

(6) Not putting a warranty on the collector covers, and supplying a cover that will need replacement if a power or equipment failure ever occurs on a sunny day.

(7) Telling the customer that it "takes a collector one-half (or one-fourth, or 15% to 30%) of the square footage of the home to heat it."

(8) Failing to warn the customer that the solar unit has not been built to conform to existing national building codes and thus might, in some cases, invalidate the homeowner's insurance when added to the house.

(9) Selling solar collectors to homeowners for daytime use only—that is, without any storage—despite the fact that a prop-

erly insulated home with the drapes open in the daytime rarely needs supplementary heat, even in far northern climes in midwinter.

(10) Selling hydronic (water-type) solar collectors that have aluminum in contact with water, without warning the consumer of the tremendous problems that may result from the formation of aluminum hydroxide.

(11) Failing to tell the potential customer that insulating the home to certain standards is, in every case, less expensive than making up for the lack of that insulation with additional solar collector and storage areas.

(12) Selling the solar collector by itself, forcing the consumer to become a solar engineer.

(13) Not telling the consumer clearly that the solar collector is the very cheapest part of the whole solar heating system. A good rule of thumb is to multiply the cost of the solar collector by four or five to obtain the cost of the whole system, installed.

(14) Not giving the consumer a good, clear warranty on the product.

(15) Publishing charts of collector efficiency, which would be different for every tilt of the collector and different in every town in the country.

(16) Publishing collector-efficiency charts that are based on noontime solar collection rather than on averages taken over the entire day.

(17) Selling solar equipment without performing heat-loss calculations on the house itself.

(18) Representing one collector as being far superior to others when, in fact, all

flat-plate collectors with the same number and kind of covers and the same insulation work the same, and vertical-vane or honeycomb-type collectors only work about 4% better.

(19) Making the utilization of solar energy seem mysterious and a high-technology science when, in fact, it is very simple and easy to understand.

(20) Failing to tell the consumer that a ground-mounted solar collector works much better (5% to 30%) than one mounted on the roof.

(21) Advertising that the system has been approved by some agency of the government, or implying the same, when, in fact, none has been.

(22) Promoting shares of stock in their concerns to ludicrously high prices by deceptive publicity.

(23) Selling solar furnaces with polystyrene foam in the collector or in the storage, despite the fact that these foams melt at 155° F.

Appendix H

I. EQUATIONS

(1) $\sin\beta = \cos\delta\cos H\cos\lambda + \sin\delta\sin\lambda$

(2) $\sin\phi = \dfrac{\cos\delta\sin H}{\cos\beta}$

(3) $D_N = \dfrac{A}{e^{\left(\frac{B}{\sin\beta}\right)}}$

(4) $D_H = D_H\sin\beta$

(5) $S_H = D_N C$

(6) $I_H = D_H + S_H$

(7) $\cos i = \cos\beta\cos\phi\sin\psi_c + \cos\psi_c\sin\beta$

(8) $D = \cos i\, D_N$

(9) $\cos k = \cos(180 - \psi_c - 2\psi_r)\sin\beta + \sin(180 - \psi_c - 2\psi_r)\cos\beta\cos\phi$

(10) $S = D_N C\left(\dfrac{1 + \cos\psi_c}{2}\right)$

(11) $S_{T1} = \tau_s S$

(12) $S_{T2} = (\tau_s)^2 S$

(13) $D_{T1} = D\tau_{i1}$

(14) $D_{T2} = D\tau_{i2}$

(15) $I_{T1} = D_{T1} + S_{T1}$

(16) $I_{T2} = D_{T2} + S_{T2}$

(17) $R_M = D_N(\cos\psi_r\sin\beta - \sin\psi_r\cos\beta\cos\phi)$

(18) $\beta_s = \tan^{-1}\left(\dfrac{\tan\beta}{\cos\phi}\right)$

(19) $f\beta = \dfrac{\sin(\psi_c + 2\psi_r - \beta_s)}{\sin(\beta_s - \psi_r)}$

(20) $F_\beta = f\beta$ for $(90 - \psi_c + 1.5\psi_r) < \beta_s < \psi_c + 2\psi_r$

(21) $F_\phi = 1$ for $\beta_s < (90 - \psi_c + 1.5\psi_r)$

(22) $F_\phi = 1 - \dfrac{1}{3}\left\{[1 + \dfrac{\cos(\psi_c + \psi_r)}{f\beta}]\dfrac{\tan\phi}{\cos\psi_r}\right\}$ for $F_\phi > 0.5$

(23) $F_\phi = \dfrac{0.75 f\beta\cos\psi_r}{\tan\phi\{\cos(\psi_c + \psi_r) + f\beta\}}$ for $F_\phi < 0.5$

(24) $R_D = \rho\, F_\beta\, F_\phi\, R_M$

(25) $R_{DT1} = R_D\tau_{k1}$

(26) $R_{DT2} = R_D\tau_{k2}$

(27) $R_S \simeq .25\rho D_N C\left\{\dfrac{\sin(\frac{\psi_c}{2})}{\sin 45}\right\}\left(\dfrac{1 + \cos\psi_r}{2}\right)$

(28) $R_{ST1} = \tau_s R_S$

(29) $R_{ST2} = (\tau_s)^2 R_S$

(30) $I_{RT1} = R_{DT1} + R_{ST1}$

(31) $I_{RT2} = R_{DT2} + R_{ST2}$

II. DEFINITIONS

A = Apparent solar irradiation at air mass zero, $Btu/ft^2 \cdot hr$ (see Table 58 for monthly values)

B = Atmospheric extinction coefficient, $1/m$ (see Table 58 for monthly values)

C = Ratio of diffuse irradiation on a horizontal surface to direct normal irradiation, dimensionless (see Table 58 for monthly values)

D = Direct solar irradiation incident on collector surface, $Btu/ft^2 \cdot hr$

D_H = Direct solar irradiation incident on a horizontal surface at sea level, $Btu/ft^2 \cdot hr$

D_N = Direct solar irradiation incident on a surface at sea level which is placed normal to the sun's rays, $Btu/ft^2 \cdot hr$

D_{T1} = Direct solar irradiation transmitted through one collector cover, $Btu/ft^2 \cdot hr$

D_{T2} = Direct solar irradiation transmitted through two collector covers, $Btu/ft^2 \cdot hr$

H = Hour angle, deg

f_β, F_β = Reflector configuration factors for solar altitude, dimensionless

F_ϕ = Reflector configuration factor for solar azimuth, dimensionless

NOTE: F_β F_ϕ is the fraction of the direct solar irradiation that is reflected onto the collector surface, accounting for the radiation reflected past the sides and over the top of the collector. It is also to be noted that some reflective surfaces (mill-finished aluminum and white porcelain, for example) may not be good specular reflectors but can be assumed to have very little scatter, which in itself will tend to be self-correcting, so that the reflections can be treated as specular.

i = Angle of incidence of sunlight on the collector, deg

I_H = Total irradiation (direct and diffuse) incident on a horizontal surface, $Btu/ft^2 \cdot hr$

I_{RT1} = Total reflected irradiation (direct reflected and diffuse reflected) transmitted through one cover of a collector, $Btu/ft^2 \cdot hr$

I_{RT2} = Total reflected irradiation (direct reflected and diffuse reflected) transmitted through two covers of a collector, $Btu/ft^2 \cdot hr$

I_{T1} = Total irradiation (direct and diffuse) transmitted through one collector cover, exclusive of reflected inputs, $Btu/ft^2 \cdot hr$

I_{T2} = Total irradiation (direct and diffuse) transmitted through two collector covers, exclusive of reflected inputs, $Btu/ft^2 \cdot hr$

k = Angle of incidence of sunlight reflected to the collector surface, deg

R_D = Direct irradiation reflected from reflector surface to collector surface, $Btu/ft^2 \cdot hr$

R_{DT1} = Reflected direct irradiation transmitted through one collector cover, $Btu/ft^2 \cdot hr$

R_{DT2} = Reflected direct irradiation

R_S = transmitted through two collector covers, Btu/ft$^2 \cdot$ hr

R_S = Diffuse (sky) radiation reflected from reflector surface to the collector surface, Btu/ft$^2 \cdot$ hr

R_{ST1} = Reflected diffuse radiation transmitted through one collector cover, Btu/ft$^2 \cdot$ hr

R_{ST2} = Reflected diffuse radiation transmitted through two collector covers, Btu/ft$^2 \cdot$ hr

S = Diffuse (sky) radiation incident on the collector surface, Btu/ft$^2 \cdot$ hr

S_H = Diffuse (sky) radiation incident on a horizontal surface, Btu/ft$^2 \cdot$ hr

S_{T1} = Diffuse (sky) radiation transmitted through one collector cover, Btu/ft$^2 \cdot$ hr

S_{T2} = Diffuse (sky) radiation transmitted through two collector covers, Btu/ft$^2 \cdot$ hr

z = Solar zenith angle, deg (Note: z is the angle that the sun's rays make with the vertical, i.e. $90° - \beta$)

β = Solar altitude, deg

δ = Declination, deg

ϕ = Solar azimuth, deg

λ = Latitude, deg

ρ = Reflectance factor of reflective panel, dimensionless (a value of 0.9 is used in this book)

τ_{i1} = Transmittance factor for direct solar irradiation through one collector cover, dimensionless (see Appendix E for τ values)

τ_{i2} = Transmittance factor for direct solar irradiation through two collector covers, dimensionless (see Appendix E for τ values)

τ_{k1} = Transmittance factor for direct solar irradiation reflected to collector and transmitted through one collector cover, dimensionless (see Appendix E for τ values)

τ_{k2} = Transmittance factor for direct solar irradiation reflected to collector and transmitted through two collector covers, dimensionless (see Appendix E for τ values)

τ_{s1} = Transmittance factor for diffuse (sky) radiation through one collector cover, dimensionless (a value of 0.778 is used in this book)

τ_{s2} = Transmittance factor for diffuse (sky) radiation through two collector covers, dimensionless (a value of 0.605 is used in this book)

ψ = Tilt of a surface, measured from the horizontal, deg

ψ_c = Tilt of the collector surface, measured from the horizontal, deg

ψ_r = Tilt of the reflective panel, measured from the horizontal, deg

Ω = Orientation angle of the collector in the horizontal plane with regard to south, deg (Note: Ω is 0° in this book; however, variances up to 30° from due south, to the east or west, will not seriously impact on the daily total collector transmitted values, other than to change the time of day of peak values)

Table 58   247

III. Total flux per hour for single cover collectors = $I_{T1} + I_{RT1}$

Total flux per hour for double cover collectors = $I_{T2} + I_{RT2}$

IV. Total flux per day for single cover collectors = $\Sigma\,(I_{T1} + I_{RT1})$

Total flux per day for double cover collectors = $\Sigma\,(I_{T2} + I_{RT2})$

V. The general equation for angle of incidence of sunlight onto a flat surface is:

$$i = \cos^{-1}\{\cos z\ \cos\psi + \sin z\ \sin\psi\cos(\phi - \Omega\,)\}$$

VI. In practice, homeowners do not change reflector angles month by month to optimize collection. If optimum reflector tilt is of interest, however, a good approximating formula is:

$$\psi_r \text{ opt} \simeq 0.667\{\tan^{-1}(\frac{\tan\beta}{\tan\phi}) - 30\}$$

where β and ϕ are taken at 11:00a.m. solar time on the date in question.

VII. Values used in this book for A, B, C, and δ were derived or extrapolated from the *A.S.H.R.A.E. Handbook and Product Directory*, American Society of Heating, Refrigerating and Air-Conditioning Engineers, New York, 1974.

Table 58   VALUES USED IN CALCULATIONS, 15TH OF MONTH

	SEP	OCT	NOV	DEC	JAN	FEB	MAR	APR	MAY	JUN	JUL	AUG
A	362	376	385	390	390	386	378	363	352	346	344	350
B	0.182	0.163	0.151	0.143	0.142	0.144	0.153	0.175	0.193	0.203	0.207	0.202
C	0.098	0.077	0.065	0.058	0.058	0.060	0.070	0.092	0.117	0.131	0.136	0.125
δ	+2.35	−8.55	−18.1	−22.75	−20.6	−12.4	−2.25	+9.6	+18.6	+22.85	+21.1	+13.75

Appendix J

PROCEDURE FOR A PRACTICAL
THREE-DAY FIELD TEST TO RATE
HOME SOLAR HEATING SYSTEMS
ACCORDING TO THEIR DELIVERY
INSIDE THE HOME OF USEFUL B.T.U.'S

INTRODUCTION

Heretofore, residential solar heating systems have been tested in place on the home over a period of years. This has been an appropriate method for the solar hobbyist who had the time and inclination to instrument his prototype system thoroughly and study it carefully.

Now, however, the residential solar heating industry has progressed to the point where systems are being manufactured and sold to homeowners on a large scale. Thus, a need has arisen for a brief test that can be performed before the homeowner begins to live with the system.

The test should provide performance information that can be compared against the manufacturer's claims (if any) and used to predict future performance.

The United States Department of Housing and Urban Development has published a supplement to "Intermediate Minimum Property Standards on Solar Heating and Domestic Hot Water Systems," and the National Bureau of Standards has also published standards for testing solar heating system components (see bibliographical references a, b and c on page 257). Although the former standards attempt to begin evaluations of solar equipment in real-world situations, the latter are laboratory tests. Their equipment, methodology and accuracy requirements make them totally impractical to take into the field to test sun-using heating systems. Indeed, as of this writing the National Bureau of Standards does not even own all the equipment necessary for testing to their own and the Housing and Urban Development standards!

In addition, the standards relate to *components* as opposed to *systems*. The synergistic effects of the components on each other when functioning in a solar heating system are beyond their scope—yet these synergistic effects are critical to the evaluation of a solar heating system. For example, the efficiency of the collector is a function of its operating temperature, which is a function of storage design and the in-service range of storage temperatures. Storage temperatures are a function of the method of delivering heat to the house as well as the amount delivered; and so on.

In response to the need for a practical testing procedure, the author (ghost writing) published "Field-Test Procedure for Rating Residential Solar Heating Systems, Based on Useful Heat Delivery," in 1975. That test called for two days of collection and three days of storage stagnation, resulting in a six-day procedure. Experience has shown that that approach was needlessly long, and so it has now been shortened. In addition, the earlier test ignored the power consumption of motors on fans or pumps, which in some cases contributes significantly to the overall heat delivery of a given system. Therefore, measurement of power consumption has been added as part of this procedure.

The following test program should be considered as an extension of and very necessary supplement to the proposed and actual governmental testing standards pertaining to components. It is a field test of solar heating systems.

The test measures the amount of solar

energy collected during one day and the amount of useful heat delivered to the house following a one-day period of storage stagnation. It may be performed on a one-run basis to verify performance claims, or as part of a series of runs to establish performance claims. The test requires approximately three days and eight to twelve man hours of labor to perform the testing, plus the time required to prepare the test report.

To perform the tests, several operating characteristics of the system, including temperatures, pressures and solar heat fluxes, have to be measured. Since the types of equipment and techniques available for these measurements are diverse, the user must be acquainted with the various types of equipment and with procedural methods if satisfactory selection and performance of instrumentation are to be assured. On page 257 is a list of equipment, costing about $1,500, that will be sufficient to test most solar heating systems.

PURPOSE

The purpose of this standard is to provide a procedure for field testing the capability of a residential solar heating system to provide useful heat on demand by the house thermostat under typical operating conditions.

SCOPE

This test program applies only to active residential solar heating systems that utilize flat-plate or vertical-vane collectors and sensible heat storage. It is limited to systems in which the transfer medium enters and leaves the collector through single inlets and outlets, respectively. The transfer medium can be either a liquid or a gas, but not a mixture of the two.

DEFINITIONS

Active Residential Solar Heating Sys-

tem: A solar heating system for heating homes, which utilizes forced circulation of the collection and distribution transfer media. As shown schematically in Figure 8, it is a system that combines the means for collecting, controlling, transporting and storing solar energy with the primary heating system in the home.

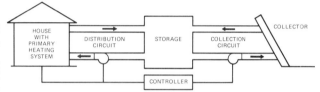

FIGURE 8: Elements of an active residential solar heating system.

Coefficient of Performance (C.O.P.): The ratio of the amount of useful energy delivered at drawdown by the solar heating system to the amount of energy used by motor(s) and other electrical equipment integral to the solar heating system during the testing period. Since the solar radiation available to the system will vary from test to test, the C.O.P. is expressed in terms of the available solar radiation.

Collection Circuit: The path followed by the collection transfer medium as it removes heat from the collector and transfers it to storage.

Collector Efficiency: The ratio of the amount of energy usefully transferred from the collector into storage to the total amount of solar energy transferred through the cover(s), expressed as a percentage.

Distribution Circuit: The path followed by the distribution transfer medium as it removes heat from storage and transfers it to the home.

Drawdown: The removal of all useful heat from storage.

Flat-Plate Collector: A device for collecting solar radiation and converting it to heat. It is stationary and does not concentrate the incoming radiation unless it is equipped with reflectors. In most embodiments, it consists of the five parts shown schematically in Figure 9:

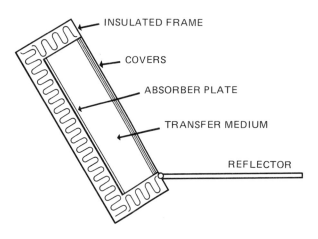

FIGURE 9: Components of a flat-plate collector.

(a) Insulated Housing: Supports the cover(s) and the absorber plate while reducing heat losses from the back and sides of the collector.

(b) Absorber Plate: Absorbs solar radiation and converts it to heat.

(c) Cover(s): Admit(s) solar radiation and insulate(s) against heat losses from the front of the collector.

(d) Transfer Medium: Carries heat away from the collector. It may be either a liquid or a gas. The medium, which can flow either in front of or behind the absorber plate, may flow through tubes thermally bonded to the absorber plate.

(e) Reflector: A mirror, polished metal or light-colored surface used to increase shortwave radiation input into the collector. It may be hinged to allow adjustment, or permanently installed at horizontal to the base of the collector, providing concentration capability of low levels to the collector.

Heat-Exchange Flow Pattern: The relative-flow arrangement of the collection and distribution circuits in storage. The typical patterns are:

(a) Counterflow: Collection and distribution transfer media flow in opposite directions through storage; used to enhance stratification.

(b) Parallel: Collection and distribution flow in the same direction through storage.

Primary Heating System: A system used to heat the house when the solar heating system can not provide all of the useful heat required to keep the house at a desired temperature, usually a conventional oil, gas or electric furnace or heat pump.

Pyranometer: A radiometer used to measure the total solar energy flux incident on a surface. This includes direct radiation from the sun, diffuse radiation from the sky, and reflected radiation (albedo) from the ground and the surroundings.

Sensible Heat Storage: A heat-storage medium in which the addition or removal of heat results in a temperature change only (as opposed to phase change, chemical reaction, etc.). The medium is typically water, water-and-antifreeze solution, bricks, rocks or fine pebbles.

Solar Furnace: A unitized, self-contained solar heating system embodying collector, reflector, pumps or fans, controls and storage in one unit.

Solarimeter: See, "pyranometer."

Stagnation: A condition in which heat is neither added to nor removed from storage mechanically, but only as a result of natural heat transfer from the storage container.

Stratification: The existence of persistent temperature gradients in storage media.

Thermal Lag: The amount of heat necessary to reachieve downpoint temperature after collection is resumed following a period of stagnation.

Useful Heat: Heat delivered by the solar heating system on demand by the thermostat, and which contributes to the reduction in the conventional heating bill. Useful heat replaces an equivalent amount of heat that otherwise would have to be provided by the primary heating system.

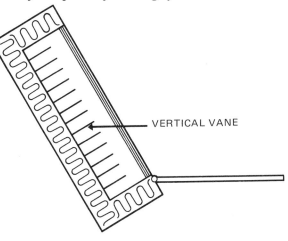

FIGURE 10: A vertical-vane collector.

Vertical-Vane Collector: Improvement to the conventional flat-plate collector by any substantial treatment of the absorber plate that results in large increases of absorber surface, designed to increase internal reflection of shortwave solar radiation transmitted through the cover(s) to reduce reflective losses from the collector. See Figure 10.

CONSTRAINTS

A residential solar heating system shall be tested in accordance with the following provisions:

(a) The system shall be field tested.

(b) The system shall be tested as a whole. If a manufacturer provides only one component of a system, such as the collector, the manufacturer's recommendations as to incorporation into the system shall be followed and the testing performed on that total system.

(c) The test series shall be performed over a period of three days, during which the first day will be devoted to collection testing, the next day to stagnation (storage) testing and the last day to drawdown testing.

(d) Collection testing shall be performed only on days when cloudiness, if any, is reasonably uniformly distributed throughout the day and the sun is not obscured by clouds for more than 50% of the daylight hours.

(e) Temperature and radiation data shall be recorded at least once every five minutes during the tests to which they apply.

(f) The accuracy of the instrumentation and measuring techniques shall be within ± 5% of the true value of the measured quantities.

(g) The test program shall be directed by a professional engineer licensed to practice in the state where testing is performed.

INSTRUMENTATION

Temperature: Measurements shall be made with either thermocouples, resistance temperature detectors (R.T.D.'s), thermistors or laboratory-grade glass-stem thermometers. Selection should be based upon the specific application and the accuracy and rate of response required of the measurement.

Flow Rate: Liquid flow rates should be measured with either a flow meter or a weight tank. Acceptable flow meters include the positive displacement, turbine and magnetic types and rotameters. Gas flow rates shall be determined by measuring the pressures and the velocities of the gas and then multiplying the average velocity by the cross-sectional area of the flow passage, care being taken that established flow patterns are being measured. The combination Pitot-static tube/inclined manometer is recommended as the simplest and most reliable technique for determining the gas velocities, particularly if the velocities are greater than 400 ft/min. Fan performance curves may be used if the flow conditions are similar to those under which the fan was tested (*e.g.*, A.M.C.A. Standard 210).

Solar Radiation: A pyranometer shall be used to measure the total shortwave radiation received from sun, sky and surroundings. Either the Eppley (Figure 11) or Moll-Gorczynski (Figure 12) type of pyranometer is recommended for this measurement.

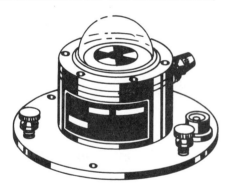

FIGURE 11: Eppley pyranometer.

FIGURE 12: Moll-Gorczynski type of pyranometer.

Power Consumption: A standard watt/hour meter shall be used to measure the energy consumed by all electrical equipment of the solar heating system.

Time: A timepiece shall be used to measure the operational time of the system's fans and/or pumps. A strip-chart event recorder may be used.

Recording Equipment: Strip-chart recorders shall be used to record temperature and pyranometric data. Manual read-out equipment is acceptable for recording flow-rate measurements.

TEST PROCEDURE

Test Preparation: The test equipment shall be set up as shown schematically in Figure 13 and the downpoint temperature shall be achieved by drawing down storage. If storage is below downpoint, this pretest drawdown will have to be preceded by a period of collection sufficient to bring storage above downpoint. The timing of this activity should be such that the drawdown is completed just before the testing is to begin. During the pretest drawdown, the storage inlet (return) temperature shall be controlled within ± 2° F of the return temperature that will be maintained during the final drawdown on Day 3. Any differences between the return temperatures during the two drawdowns will be used to correct for the differing amounts of heat remaining in storage following each drawdown.

Collection Testing (Day 1): The test series shall begin with one full day of collection, starting at the downpoint temperature of storage. Heat is not to be removed from storage by distribution during this period. However, it is not appropriate to provide special sealing of distribution ducts or transfer areas because, to maintain the real-world conditions, the solar heating system equipment should have provisions for stop-

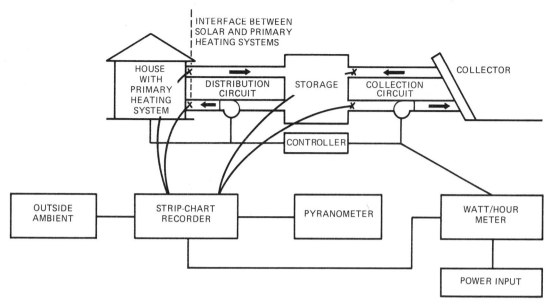

FIGURE 13: Diagram showing the test equipment setup.

ping heat distribution during collection when heat is not required by the residence. The following data shall be recorded during collection testing:

(a) Differential temperature of the collection transfer medium across storage: This shall be determined by placing temperature probes in the storage inlet and outlet openings in the collection circuit.

(b) Flow rate of the transfer medium: Measurements shall be made at some point in the collection circuit where the temperature of the transfer medium is known, preferably at one of the locations specified in (a) above.

(c) Net collector area: Measure the area of the collector cover, less the area shaded by glazing and glazing supports.

(d) Time of collector fan or pump operation: This requires measuring the cumulative time that the collector pump or fan operates.

(e) Power consumption of the solar furnace during collection: This requires measurement of the cumulative power consumption by all electrical equipment inte-

gral to the solar heating system during collection.

(f) Amount of shortwave radiation incident on the collector: The pyranometer shall be located horizontally near the collector, but not in a position where any reflectors, if used, will reflect light on it.

(g) Meteorological data: The outside ambient temperature shall be measured with a temperature probe housed in a well-ventilated instrumentation shelter designed to U.S. Weather Bureau specifications. In addition, periodic observations of wind speed and direction should be made. Barometric pressure readings must be taken if the test results are to be converted to standard atmospheric conditions.

Stagnation Storage Testing (Day 2): During this 24-hour period, no collection or distribution is permitted. The only data recorded during this period is the air temperature surrounding the storage container, or the ground temperature surrounding the storage container if it is located underground. Periodic observations of the wind speed and direction should be made if the storage container is outside and above ground.

Drawdown Testing (Day 3): During this test, the following data shall be recorded:

(a) Differential temperature of the distribution transfer medium across storage: This shall be determined by placing temperature probes in the distribution circuit at the interface between the solar and the primary heating systems.

(b) Temperature of the distribution transfer medium at the inlet to storage: This temperature, which shall be maintained throughout drawdown at a predetermined temperature ± 2° F, shall be measured by the return probe(s) in (a) above.

(c) Flow rate of distribution transfer medium: These measurements shall be made at some point in the distribution circuit where the temperature of the transfer medium is known, preferably at one of the locations specified in (a) above.

(d) Time of distribution fan or pump operation: This requires the recording of the time elapsed during complete drawdown.

(e) Power consumption during distribution: This requires measurement of the electrical power consumed by the solar unit during drawdown.

TEST RESULTS

Using the data recorded as above, the following performance information shall be compiled:

Collection Capability: Cloudless-day heat transfer to storage of _____ Btu's/day on __(date)__ with an average daily shortwave radiation of _____ Btu's/ft on a horizontal surface and an initial storage temperature of _____°F.

Storage Capability: Delivery of useful heat of_____% of the energy stored before a 24-hour period of stagnation, during which

the average air (or ground) temperature surrounding the storage container was_____°F and average wind speeds were_____mph.

Average Collector Efficiency: With an *average* of_____ Btu/ft²·hr transmitted through the collector cover(s), the average collector efficiency was_____% with an average outside temperature of_____°F.

Coefficient of Performance during Test: Tested C.O.P. was_____with total available solar radiation of_____ Btu/ft²·day on a horizontal plane.

CALCULATION OF TEST RESULTS

These shall be calculated as follows:

(a) Compute the total heat input to storage during the day of collection (Q_s).

$$Q_s = \int \dot{m}\, c_p\, \Delta T\, dt$$

(b) Compute the average daily solar radiation as measured by the pyranometer during the day of collection (G_i).

$$G_i = \int \phi_i\, dt$$

(c) From Figure 14, determine the cloudless-day solar radiation to the radiation on the test day (k).

the test (G_t).

(d) Compute the ratio of cloudless-day solar radiation to the radiation on the test day (k).

$$k = G_t/G_i$$

(e) If fans or pumps are located in such a fashion as to conserve thermal losses from the motors (by having them in the fluid stream), calculate the thermal losses from the collector motors (L_c) as follows:

$$L_c = 0.3P_c$$

after converting measured power consumption from watt · hours/day to Btu/day.

(f) Calculate the cloudless-day heat transfer from the collector to storage (Q_a).

$$Q_a = k(Q_s - L_c)$$

(g) From the table applicable to the type of collector (from Tables 37-44), find the total Btu's transmitted to the inside of

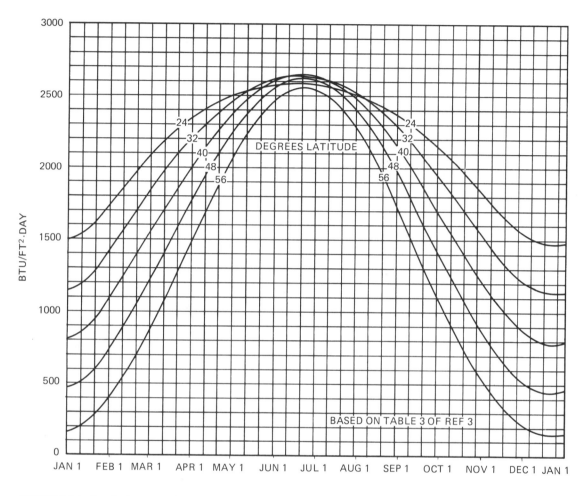

FIGURE 14: Graph shows total cloudless-day solar radiation on a horizontal surface.

the collector at the test location (R_t). If collector tilt is not 60°, calculate values.

(h) From Table 45, find the possible hours of sunshine for the date of the test at the test location (S).

(i) Calculate the average radiation transmitted to the inside of the collector on the test day.

$$\phi_a = \frac{Rt}{kS}$$

(j) Compute the average collector efficiency for the test period (E_c).

$$E_c = 100\ \left(\frac{Q_a}{R_t\ A_c}\right)$$

(k) If the distribution fans or pumps are located in such a fashion as to conserve thermal losses from the motor(s) by having it (them) located in the fluid stream, calculate the thermal losses from the distribution motor(s) (L_d) as follows:

$$L_d = 0.3\ P_d$$

after converting measured power consumption from watt · hours/day to Btu/day.

(l) Compute the total heat delivered to the house (Q_h).

$$Q_h = \int \dot{m}\, c_p\, \Delta T\, dt$$

(m) Calculate the total useful heat delivered to the house (Q_d).

$$Q_d = Q_h - (L_d + L_c)$$

(n) Calculate the percentage of stored heat delivered after 24 hours of stagnation (E_s).

$$E_s = 100\left(\frac{Q_d \pm Q_r}{Q_s}\right)$$

where Q_r is the difference in the amount of stored heat before and after the three-day test. Q_r will have a nonzero value if the storage inlet temperatures differ between the pretest and the final drawdowns.

(o) Calculate the Coefficient of Performance (C.O.P.).

$$COP = \frac{Q_h}{P_c + P_d}$$

TEST REPORT

A new report should be written and submitted to the person or agency contracting for the test, upon conclusion of the testing. The report should include:

(a) Manufacturer's name and address.

(b) Product name (trade name) and model number.

(c) Testing agency's name and address.

(d) Dates and location of the test series.

(e) Test description, including a description of the system (article) tested, the instrumentation used and the procedure followed. A detailed time-line of the test series should also be provided.

(f) Test results consisting of at least one copy of:

(i) The raw data, including strip-chart read-outs.

(ii) The reduced data, with appropriate remarks: *e.g.*, weather conditions during collection and stagnation testing.

(iii) The performance information required in the Test Results section above, including the calculations that followed.

(g) If requested, an engineering evaluation of the over-all system, including:

(i) An estimate of the probable service life, taking into consideration such factors as wear, weathering, accidental damage, etc.

(ii) An estimate of the annual operating costs (power requirements, repairs, maintenance, lubrication, labor, etc.)

(iii) A determination of the product's conformity to safety standards.

In addition, the report should bear the seal or facsimile and signature of the registered professional engineer responsible for the test.

NOMENCLATURE

c_p — specific heat of transfer medium at constant pressure, Btu/lb·°F

d_t — time derivative, hr

E_c — collector efficiency as a ratio of solar energy transmitted through covers to energy usefully transferred to storage, %

E_s — storage efficiency as ratio of energy transferred to storage to energy usefully transferred from storage after 24 hour stagnation period, %

G_i — average daily solar radiation measured at the test site, Btu/ft^2· day

G_t — cloudless-day solar radiation, Btu/ft^2· day

k — ratio of cloudless-day radiation to test-day radiation, dimensionless

L_c — thermal loss from collector motor, Btu/day

L_d — thermal loss from distribution motor, Btu/day

m — mass flow of transfer medium, lb/hr

Q_a — cloudless-day heat transfer to storage after subtraction of electrical heat inputs, if any, Btu

Q_d — net heat delivered to the residence, after subtracting electrical heat inputs, if any, Btu

Q_h — total heat delivered to the house, Btu

Q_r — residual heat in storage between the start and end of the test, Btu

Q_s — total heat input to storage, Btu

R_t — solar energy transmitted through cover(s) of collector, Btu/ft^2· day

ΔT — differential temperature, °F

ϕ_a — average radiation transmitted to inside of collector, Btu/ft^2· hr

ϕ_i — incident solar radiation, Btu/ft^2· hr

S — possible sunshine, hr/day

A_c — area of collector, ft^2

REFERENCES

(a) *Intermediate Minimum Property Standards Supplement, Solar Heating and Domestic Hot Water Systems*, U.S. Department of Housing and Urban Development, 1977.

(b) *Method of Testing for Rating Thermal Storage Devices Based on Thermal Performance*, National Bureau of Standards Interim Report 74-634, December 1974.

(c) *Method of Testing for Rating Solar Collectors Based on Thermal Performance*, National Bureau of Standards Interim Report 74-635, December 1974.

(d) "Solar Energy Utilization for Heating and Cooling," *A.S.H.R.A.E. 1974 Applications Handbook*, American Society of Heating, Refrigerating and Air-Conditioning Engineers, New York, 1974.

(e) *Harnessing the Sun*, Keyes, J.H., Morgan & Morgan, Publishers, Dobbs Ferry, New York, 1975.

(f) *Field Test Procedure for Rating Residential Solar Heating Systems Based on Useful Heat Delivery*, ISC-01-23, I.S.C. Engineering Division, Nederland, Colorado, 1975.

REPRESENTATIVE LIST OF INSTRUMENTATION

Item	Estimated Cost
Watt/hour meter	$ 100
Moll-Gorczynski type of pyranometer	350
Strip-chart recorder	850
Flow-measuring equipment	150
Misc.	50
Total	$1,500

NOTE CONCERNING ACCURACY OF TEST MEASUREMENTS AND CALCULATIONS

The nature of this test as a field test renders it impractical to obtain accuracies greater than ± 5%. In addition, the uses to which the test results will be put do not warrant greater accuracy. A dwelling's heating requirements cannot be determined practically within ± 5% accuracy, nor can the solar energy available at most test sites. Calculating cloudless-day collection on the ratio of pyranometric data to theoretical solar radiation is an approximation, as are the calculations of collector efficiency. Distribution of cloud cover during the day will affect the accuracy of the calculations. This is not significant within the tolerances of the test, except when the majority of the permissible 50% cloud obstruction occurs around solar noon.

NOTES ON TEST PROCEDURE

Basis for Testing to the Downpoint Temperature: Since it is uneconomical to design a solar heating system to provide 100% of the heating requirements of a house, it follows that all the solar heat collected will be used in typical midwinter operation. Thus, in midwinter, collection and storage temperatures will be controlled by the system's downpoint temperature. To test for this condition requires starting and ending the test at the downpoint temperature. Typical downpoint temperatures are 75° F for forced-air/pebble-bed systems and 160° F for all-hydronic systems.

Basis for Using Collection and Distribution Transfer Media Rather Than Storage Temperatures, to Determine the Amount of Heat Transferred: Measuring storage temperatures is difficult in most cases because it is difficult to instrument storage properly after the system has already been installed, and the stratification effects inherent in certain storage media (*e.g.*, pebble beds) require the use of many probes to determine the average storage temperature at any one time.

In addition, when the transfer medium temperatures are measured at the location shown in Figure 13, the heat losses in the collection and distribution circuits are included in the calculations—as they should be in a *system* test.

Basis for the 24-Hour Stagnation Period: To be effective, a solar heat storage unit must have the capability of supplying heat during noncollection periods. One day was selected, since it represents an average requirement of a system supplying less than 100% of a home's heating requirements. It is long enough to allow for measurable heat flow, but not too long for convenient field testing.

Appendix K MISCELLANEOUS TABLES

Table 59 STASIS TEMPERATURE ADDITIVES

Instructions: To determine temperature inside a given collector under stasis (no-flow) conditions, add outside temperature to the appropriate additive from the tables below. Refer to Tables 60 and 61 for $Btu/ft^2 \cdot hr$ transmitted at solar noon on 15th of each month.

SINGLE-COVER COLLECTORS

Btu/ft^2 Transmitted Through Cover	Glass-Fiber Batt Insulation				Urethane Foam	
	R-3.33, 1"	R-6.67, 2"	R-8.33, 2-1/2"	R-10, 3"	R-15, 2"	R-30, 4"
100	17	18	19	19	19	20
125	29	31	32	32	33	33
150	40	44	45	45	46	47
175	51	56	57	57	58	60
200	63	68	69	69	71	72
220	73	79	80	81	83	84
250	84	90	92	93	94	96
275	94	101	103	104	106	107
300	104	112	113	115	117	118
325	114	122	124	125	127	129
350	124	132	134	135	138	140
375	133	142	144	145	148	150
400	142	152	154	155	158	160
425	151	161	163	165	167	170
450	160	171	173	174	177	179

DOUBLE-COVER COLLECTORS

Btu/ft^2 Transmitted Through Covers	Glass-Fiber Batt Insulation				Urethane Foam	
	R-3.33, 1"	R-6.67, 2"	R-8.33, 2-1/2"	R-10, 3"	R-15, 2"	R-30, 4"
75	16	20	20	20	21	22
100	35	40	41	42	44	45
125	53	60	62	63	65	67
150	70	79	81	82	85	87
175	86	97	99	101	103	106
200	102	113	116	118	121	124
225	117	129	132	134	138	141
250	131	145	148	150	153	157
275	145	159	162	165	168	172
300	158	173	177	179	183	187
325	171	187	190	192	196	200
350	183	199	203	205	209	214
375	195	212	215	218	222	226
400	206	224	227	230	234	238

Notes to Table 59: *(1) Fiber glass—k = 0.3 $Btu/ft^2 \cdot hr \cdot in.$; polyurethane foam—k = 0.133 $Btu/ft^2 \cdot hr \cdot in.$; single cover—1.13 $Btu/ft^2 \cdot hr$; double cover—0.45 $Btu/ft^2 \cdot hr$. (2) Stefan Boltzmann constant $=0.173 \times 10^{-8}$ $Btu/ft^2 \cdot hr \cdot {}^\circ R^4$. (3) Cover emissivity = 0.83. (4) Mean sky and surroundings temperature assumed for purposes of these tables to be same as outside ambient temperature. (5) Comment: These calculated values fall as much as 23% below heuristic values for vertical-vane collectors; hence, recommend 400°F design temperatures for collectors.*

Table 60 CLOUDLESS B.T.U./FT²·HR TRANSMITTED THROUGH COVER(S) AT SOLAR NOON ON 15TH OF MONTH

Degrees North Latitude	Sep	Oct	Nov	Dec	Jan	Feb	Mar	Apr	May	Jun	Jul	Aug
60° TILT SINGLE COVER WITHOUT REFLECTOR												
25	221	255	272	279	279	270	245	200	161	142	148	179
30	233	261	273	276	278	275	256	216	181	162	168	195
35	242	264	270	269	273	276	263	229	197	181	185	210
40	248	264	263	258	264	274	267	240	212	197	201	222
45	251	260	250	240	249	267	268	247	224	210	213	231
50	251	252	231	212	226	256	266	252	234	222	224	237
55	247	238	203	171	192	238	259	253	240	230	231	240
60° TILT SINGLE COVER WITH REFLECTOR												
25	224	259	307	335	325	283	248	203	165	147	153	183
30	237	278	329	355	348	307	259	219	185	167	173	200
35	246	299	348	367	363	330	275	233	202	185	190	214
40	262	320	359	354	368	350	298	243	216	201	205	226
45	282	336	343	317	335	363	321	256	228	215	218	235
50	301	346	305	269	292	353	339	275	238	226	228	247
55	316	325	255	205	236	316	351	295	250	235	237	265
60° TILT DOUBLE COVER WITHOUT REFLECTOR												
25	191	212	240	246	246	237	213	172	137	120	125	154
30	202	229	240	243	245	242	223	187	155	138	144	168
35	211	232	238	237	240	243	231	198	170	155	159	181
40	217	232	231	227	233	241	235	208	183	169	173	192
45	220	229	220	211	219	235	236	216	194	182	184	200
50	220	221	203	187	199	225	234	221	203	192	193	207
55	217	210	179	151	169	210	228	222	209	200	200	210
60° TILT DOUBLE COVER WITH REFLECTOR												
25	194	215	263	288	279	245	215	175	141	123	129	157
30	205	240	283	306	300	263	225	189	159	142	148	172
35	214	257	300	319	314	283	239	201	173	159	163	184
40	226	275	312	309	321	302	256	211	186	173	176	195
45	241	290	299	278	293	315	275	221	197	185	188	203
50	258	300	267	236	256	308	292	236	206	195	197	213
55	272	284	224	180	207	277	304	252	216	203	205	227

Note to Tables 60 and 61: For Ground-Mounted Units, Add 20 %

Table 61   261

Table 61 CLOUDLESS B.T.U./FT²·HR TRANSMITTED THROUGH COVER(S) AT SOLAR NOON ON 15TH OF MONTH

Degrees North Latitude	Sep	Oct	Nov	Dec	Jan	Feb	Mar	Apr	May	Jun	Jul	Aug
90° TILT SINGLE COVER WITHOUT REFLECTOR												
25	93	147	189	207	201	169	119	56	21	15	17	39
30	116	168	203	218	214	188	143	83	39	24	31	60
35	138	184	214	226	223	204	164	106	61	43	50	85
40	157	198	221	227	228	215	183	129	86	64	73	106
45	173	207	221	221	225	223	198	149	108	88	95	128
50	185	212	214	206	214	225	209	168	129	110	117	146
55	194	211	196	173	191	219	216	182	148	130	135	162
90° TILT SINGLE COVER WITH REFLECTOR												
25	269	276	355	373	375	318	223	253	188	165	171	217
30	276	314	365	366	371	355	269	267	220	189	199	244
35	281	346	357	353	360	369	309	275	247	221	231	256
40	282	352	343	332	343	363	344	282	259	245	248	263
45	323	343	322	303	318	350	359	284	267	256	256	269
50	333	328	292	264	283	330	352	314	272	262	263	271
55	325	305	251	209	236	302	339	338	275	267	265	301
90° TILT DOUBLE COVER WITHOUT REFLECTOR												
25	74	125	164	180	174	146	99	41	15	11	13	27
30	96	144	177	190	186	163	122	65	27	17	21	44
35	116	159	186	196	194	177	141	86	45	30	36	67
40	134	171	192	198	199	187	158	109	68	48	56	87
45	149	180	193	195	198	194	171	127	89	70	76	108
50	160	184	188	181	189	196	181	145	109	90	97	124
55	168	184	173	153	168	193	187	157	126	110	115	139
90° TILT DOUBLE COVER WITH REFLECTOR												
25	228	235	308	324	325	274	185	212	160	141	146	182
30	237	271	317	319	322	307	228	225	184	160	167	205
35	243	300	311	307	313	321	266	235	208	185	193	216
40	245	305	299	291	299	316	298	244	219	205	208	225
45	280	298	281	267	279	305	311	246	228	216	217	232
50	289	285	257	233	250	289	306	271	235	224	226	234
55	282	267	221	184	208	265	295	293	236	230	229	260

Note to Tables 60 and 61: *Adapted from* Tables of Monthly Terrestrial Solar Radiation and Transmissivity, Single- and Double-Cover Collectors at 60° and 90° Tilts, With and Without Reflectors, *Keyes, J.H., Morgan & Morgan, Dobbs Ferry, N.Y., 1979.*

Appendix L

RULES FOR PURCHASING SOLAR
EQUIPMENT — A CHECKLIST

Rule Number One: If insulated alike, all flat-plate collectors with the same number and kind of covers work the same when new. Vertical vanes in flat-plate collectors will improve performance only about 4% when new. (Page 7)

Rule Number Two: A decently designed collector should have at least as much insulation on its sides and back as you have in the attic of your house. (Page 10)

Rule Number Three: Unless you live at a latitude of less than 30 degrees North, don't buy a flat-plate collector with fewer than two covers. The covers should be of tempered glass. If glass fiber or plastic is used for a cover, it should be the outer cover only, with the under cover of tempered glass. (Page 12)

Rule Number Four: Don't buy a collector that has two glass covers bonded rigidly to each other. (Page 13)

Rule Number Five: Ask about the method used by the manufacturer to prevent moisture build-up between the collector covers. The evaluate that method with your own common sense. (Page 13)

Rule Number Six: Insist upon seeing the results of a seven-day stagnation test performed by an independent testing laboratory on the collector you plan to buy. (Sometimes this is called a "stasis test.") (Page 14)

Rule Number Seven: Be sure to find out what kind of sealants have been used on the collector. Then look at the collector yourself to see just how difficult it will be to recaulk the covers after a year or so. (Page 14)

Rule Number Eight: Make sure that the manufacturer has addressed the dust problem. Be very sure that the buck hasn't been passed to you. (Page 15)

Rule Number Nine: Unless you have money in great abundance, and an all-consuming desire to be parted from it foolishly, you will purchase a forced-air solar heating system for your home. This is true even if you are adding solar heat to a home that already has hot-water heat. (Page 21)

Rule Number Ten: Don't buy a solar heating system unless it has a warranted nameplate Btu Output Rating telling you what the Btu output from the entire system is. (Page 23)

Rule Number Eleven: Never buy anything—not even a cup of coffee—from anyone who tells you, "It takes a collector one-half (or one-fourth, or 15% to 30%) the square footage of your home to heat it." Run —don't walk—away from that showroom. Worse yet, if that kind of statement appears in the manufacturer's literature, you are in bad company. Don't buy. (Page 23)

Rule Number Twelve: The *only* way you can size a solar furnace properly to your home is to perform a heat-loss calculation on your home and to know the *useful* Btu output of the solar unit in question. (Page 24)

Rule Number Thirteen: Do not, under any conditions, purchase a solar/water collector that has aluminum in contact with water. Demand copper. (Page 24)

Rule Number Fourteen: Be sure to find out how much antifreeze the system holds, how much that antifreeze costs and how often it must be changed. (Page 25)

Rule Number Fifteen: Unless you are stuck with putting the solar collector on the roof, don't buy a solar furnace without reflective panels. (Page 26)

Rule Number Sixteen: Do not buy a concentrating collector solely to heat your home. The *only* justification for using a concentrating collector is to obtain higher temperatures not needed for home heating. (Page 28)

Rule Number Seventeen: Do not purchase a concentrating collector for home heating or cooling without seeing a warranted Btu Output Rating for the *entire* system, inside the house, under real-world conditions. Compare that with the rating for flat-plate collectors delivering the same output and compare total costs. (Page 30)

Rule Number Eighteen: Don't be taken in by talk about how long the storage unit will store heat, particularly if the time periods talked about are more than a day-and-a-half. (Page 32)

Rule Number Nineteen: Don't buy a solar furnace where the storage has less insulation than you have in your attic, if it is to be located outside or below ground (R-30 minimum). If the storage is to be located inside the house, don't accept less than R-15 insulation. (Page 34)

Rule Number Twenty: If you have any choice, insist upon pebble-bed storage.

(Page 37)

Rule Number Twenty-one: If the solar heating system does not have a proportional controller—one that compares temperatures in storage and on the collector to determine when to turn on the collector pump or fan—don't buy the system. (Page 42)

Rule Number Twenty-two: Do not buy a solar hot-water heater with a water-type collector unless you have seen a certificate of compliance with all building codes. (Page 43)

Rule Number Twenty-three: It *always* costs less to upgrade your home's insulation to the necessary standards than to buy additional solar collector equipment to make up for not doing so. (Page 48)

Rule Number Twenty-four: No matter how large the manufacturer is, never buy components. Insist upon buying a complete, pre-engineered solar heating system with a warranted system output. (Page 62)

Rule Number Twenty-five: Unless you got rich trading in penny uranium stocks, don't dabble in solar stocks, at least until 1980. (Page 64)

Rule Number Twenty-six: Do not even shop for solar heating equipment until you have determined the Degree-Day size of your house. (Page 66)

INDEX

41663